FEMINIST TECHNOSCIENCES
Rebecca Herzig and Banu Subramaniam, Series Editors

HOLY SCIENCE

THE BIOPOLITICS OF HINDU NATIONALISM

BANU SUBRAMANIAM

UNIVERSITY OF WASHINGTON PRESS
Seattle

Financial support for the publication of *Holy Science* was provided by the Office of the Vice Chancellor for Research and Engagement, University of Massachusetts Amherst.

Interior design by Katrina Noble
Composed in Iowan Old Style, typeface designed by John Downer

23 22 21 20 19   5 4 3 2 1

UNIVERSITY OF WASHINGTON PRESS
www.washington.edu/uwpress

LIBRARY OF CONGRESS CATALOGING-IN-PUBLICATION DATA
Names: Subramaniam, Banu, 1966- author.
Title: Holy science : the biopolitics of Hindu nationalism / Banu Subramaniam.
Description: Seattle : University of Washington Press, [2019] | Series: Feminist
    technosciences | Includes bibliographical references and index. |
Identifiers: LCCN 2018048245 (print) | LCCN 2019002318 (ebook) |
    ISBN 9780295745602 (ebook) | ISBN 9780295745589 (hardcover : alk. paper) |
    ISBN 9780295745596 (pbk. : alk. paper)
Subjects: LCSH: Biopolitics—India. | Hinduism and science—India. | Hinduism
    and politics—India. | Nationalism and science—India. | Postcolonialism—India.
    | India—Politics and government—1947
Classification: LCC JA80 (ebook) | LCC JA80 .S87 2019 (print) |
    DDC 320.540954—dc23
LC record available at https://lccn.loc.gov/2018048245

Chapter 1 appeared in a different form as "Development Nationalism: Science, Religion, and the Quest for a Modern India," in *Feminist Futures: Re-Imagining Women, Culture and Development*, edited by Priya Kurian, Kum-Kum Bhavnani, and John Foran (London: Zed Books, 2016).

Cover illustration: Based on image of holy *trishul*, DigitalSoul/iStock

The paper used in this publication is acid free and meets the minimum requirements of American National Standard for Information Sciences—Permanence of Paper for Printed Library Materials, ANSI Z39.48–1984.∞

Caminante, no hay puentes, se hace puentes al andar.

Voyager, there are no bridges, one builds them as one walks.

GLORIA E. ANZALDÚA, *This Bridge Called My Back*

हाँ कहूँ तोई है नाही
ना भी कहियो नाही जावे
हाँ और ना के बीच मे
मोरा सदगुरु रहा समाये

If I say Yes, it isn't so
And yet I cannot say it's No
That's where my guru resides
Between that Yes and No.

KABIR

CONTENTS

PROLOGUE. In Search of India: The Inner Lives
of Postcolonialism   ix

Avatars for Lost Dreams: The Land of Lost Dreams   xv

INTRODUCTION. Avatars for Bionationalism: Tales from (An)Other
Enlightenment   3

Avatar #1: The Story of Uruvam   46

1. Home and the World: The Modern Lives of the Vedic Sciences   49

Avatar #2: The Story of Amudha   72

2. Colonial Legacies, Postcolonial Biologies: The Queer Politics of
(Un)Natural Sex   76

Avatars #3: The Story of Nādu and Piravi   108

3. Return of the Native: Nation, Nature, and Postcolonial
Environmentalism   113

Avatar #4: The Story of Néram   141

4. Biocitizenship in Neoliberal Times: On the Making of the "Indian"
Genome   145

Avatar #5: The Story of Arul   178

5. Conceiving a Hindu Nation: (Re)Making the Indian Womb   182

Avatar #6: The Story of Kalakalappu   206

CONCLUSION. Avatars for Dreamers: Narrative's Seductive
   Embrace   209

   Notes on the Mythopoeia   215

EPILOGUE. Finding India: The Afterlives of Colonialism   223

   A Note of Gratitude and Appreciation   231

   Notes   237

   References   243

   Index   285

PROLOGUE

In Search of India: The Inner Lives of Postcolonialism

Nations are built on wishful versions of their origins: stories in which our forefathers were giants, of one kind or another. This is how we live in the world: romancing. . . . When we remember—as psychologists so often tell us—we don't reproduce the past, we create it.

HILARY MANTEL, "Why I Became a Historical Novelist"

How to stop a story that is always being told? Or, how to change a story that is always being told?

AUDRA SIMPSON, *Mohawk Interruptus*

THIS IS AN ODE TO INDIA. THE INDIA THAT I WAS BORN INTO, THE India I grew up in, the India that breathed into me life, hope, passion, and imagination. India suffuses the senses. The vibrant sounds of India, the blaring Bollywood music, the morning calls of the mosque, the rhythmic chants of the priest, the church bells, the neighbor's pressure cooker, the barking dogs, the car horns—there is music in this cacophony. The arresting sights of colors in the bazaar, the vibrant saris, the movie posters, gaudy cutouts of politicians, temples in every corner, the blue skies, the scorching sun, and the shade of black umbrellas. Smells overpower the senses, the fragrance of jasmine, the aroma of brewing tea, the drying patties of cow manure, the neighbor's redolent fish curry, the scent of milk boiling. And always, the intoxicating smell of the earth as the monsoons arrive. The sights, smells, and sounds of India reverberate through these pages.

As I delve into India, a cornucopia of riches awaits, endlessly complex and filled with knotty contradictions. The answers lie in contradictions, in incommensurability, in the silences, in love and hate, in the infinite spaces between Kabir's zero and one, within India's syncretic traditions, communal tensions, friendships, rich arts, supple sciences, Hindu supremacy, secular hopes, and the vibrant religions and philosophies that have thrived in this region. Perhaps we should start at the beginning, on the eve of India's independence. . . .

In the evocative words of India's first prime minister, Jawaharlal Nehru, at the stroke of midnight on August 14, 1947, while the world slept, India awoke to life and freedom and began its tryst with destiny.[1] Approaching its seventieth year of independence in 2017, one commentator remarked that India "seems to have missed its appointment with history" (P. Mishra 2017). Why? In 1947, after centuries of migrations, invasions, conquest, and colonial rule, the rich, heterogeneous, and fractured geographies of South Asia were partitioned into what became the Sovereign Republic of India[2] and the Islamic Republic of Pakistan, accompanied by unspeakable violence. With the partition began the project of consolidating "India" as an independent postcolonial nation, the largest democracy in the world: a nation marked by difference, heterogeneous histories, diverse cultures, and many religions, languages, cuisines, and contested borders. The motto Unity in Diversity signaled that in the multiplicity and pluralism of India's religions, languages, cultures, and peoples lay the central idea and project of "national integration."

This was the rousing rhetoric that many of us grew up with in the postcolony. Passionate speeches by politicians, emotional and uplifting plots in Bollywood movies, diverse languages on television, government-sponsored posters and slogans, and rousing songs and pledges in school all attempted to cajole and consolidate a singular nation. The young independent India imagined itself a modern nation, putting its hopes in science, technology, modernization, and industrialization. No doubt the rousing rhetoric attempted to soothe the countless inequalities and injuries that continued to plague the nation. In everyday life, we were witness to rampant caste, class, gender, and religious discrimination. While the state invested in grand visions of science and technology, life on the ground hummed blissfully in the familiar registers of local religious and spiritual traditions. In this ethos of postcolonial India, I set out from India to the United States to pursue a graduate education in the biological sciences. Three decades later, my travels to the West have taken me back to the East. In the meanwhile, the somewhat secular ethos of India has been thoroughly supplanted by a fervent Hindu nationalism. In 2014, for the first time in its history, a Hindu nationalist government emerged as the majority party in government.[3] Introduced to postcolonial studies, feminist studies, and science and technology studies, I now return to India with newfound interdisciplinary skills. It is clear that what has emerged in contemporary India is possible only because of the complex politics, the deep heterogeneity, the conflicting narratives, and the profound contradictions that were always woven intricately into the political fabric of India.

Under calls for an authentic "Hindu India," other genealogies lurk. Hindu nationalists invoke not only an ancient mythological corpus but also a "mythoscientific" one, challenging the very basis of academic knowing and knowledge. The "facts" being contested are vast, as our very conceptions of history, geography, politics, language, culture, biology, geology, astronomy, physics, chemistry, medicine, engineering, and mathematics are now contested terrain. How do we reconcile these? What is India? Who gets to define it? Which India do we embrace?

Home once interrogated, Kamala Viswesaran (1994: 113) notes, "is a place we've never been before." I thought I was a child of the enlightenment, captured, enraptured, and firmly converted to enlightenment logic and the rationality of science and technology. Yet, having moved to the West and been introduced to postcolonial and decolonial thought, I have come to realize that the liberal ideologies of modernity cloud my memories and narratives of my childhood and of India.

In thinking back to very early in my childhood, scenes and images come flashing by. As I recount these stories, I can see that my childhood memories are themselves haunted by a scientized and liberal postcolonial narrator who had learned to firmly demarcate the modern from the nonmodern. I remember my first sight of a *kavadi* procession while visiting my grandparents in rural Tamil Nadu. Colorful religious processions with gyrating bodies in a frenetic trancelike state passed through the streets.

Along with the rhythmic drumbeat, each person danced carrying an elaborate *kavadi* (burden), usually on behalf of a loved one who needed healing. Kavadi Attam ("burden dance") is a ceremonial sacrifice practiced by devotees of the god Murugan. They often pierce their skin, tongue, or cheeks with skewers. In a trance, their "possessed" bodies dance through the streets. Sometimes devotees engage in self-flagellation and walk on fire. I remember watching the procession with fear, fascination, and utter incomprehension. Similarly, during the festival of Muharram, I watched men of the Shia community self-flagellate, beating their breasts and faces, often with chains and whips, as they danced by in a trancelike state. Could human biologies endure such pain? When I inquired, I was told that these were acts of extreme devotion, ecstasy, and passion, which could overcome bodily thresholds of pain and injury. This is what true devotion is made of.

I remember family visits to holy temples and shrines whose inner walls bore erotic sculptures of voluptuous, undulating, entangled bodies, genitals casually displayed in open daylight. Surrounding us on the walls were figures of multiple genders seeming to copulate in various configurations and postures. The Christian and Sufi saints, Hindu *sants*, Muslim *pirs*, and syncretic sites all across the country beckoned and healed. In the inner sanctum of Shiva temples, the *lingam* rises from the ground in the form of a phallus, often emerging from inside the *yoni*, or womb, a symbol of Shakti.

In books, movies, and newspaper articles, I read stories of humans marrying trees and animals, and new mothers breastfeeding orphaned deer alongside their own children. Innumerable temples are dedicated to animals regarded as gods—elephants, snakes, monkeys, and rats. Increasingly, a temple marks every corner. The walls of buildings are plastered with pictures of gods, most often a practical solution to male public urination. The gurus and godmen of India reign supreme. There is no business in India today like the god business. This multiplicity and polyvocality of India defies easy binaries of science and religion, nature and culture, human and nonhuman, modern and ancient, sacred and profane. There is nothing to "catch up" to in the West. India dances to its own drummer, often in multiple rhythms.

In my last book, *Ghost Stories for Darwin*, I explored the history of science and the tremendous damage that eugenics has wrought on the world.

The "eugenic ghosts" remind us about the power of science and about the lives of the ghostly dispossessed—the bodies violated, sterilized, experimented upon, maimed, killed, and exterminated. But in exploring the history of biology, I was also struck by the vibrancy of scientific thought. What was always apparent is that at every moment of history, other ideas existed—a strong reminder that nothing is inevitable, and other lives and other futures were and are always possible.

Contemporary India challenges us as it professes another modernity, another enlightenment. This book is about such tales, and the elisions and erasures in the logics of "another" world. Yet one must ask, if we reject colonialism, must we get religious nationalism? The answer is decidedly no. This book is dedicated to the belief that we do not have to choose between binary logics. We can instead embrace science and religion, nature and culture, human and nonhuman to imagine worlds that defy imperial Western logics and nativist religious nationalisms.

AVATARS FOR LOST DREAMS

The Land of Lost Dreams

Planet Kari is entering a new century, and feverish activity envelops the nearby Avatara Lokam, or Land of Lost Dreams. The Avatars in the Land of Lost Dreams watch nervously as life unfolds on Kari. Destruction, degradation, and decay are everywhere on Kari. The doomsday callers sound their alarms. The lands, oceans, and skies are heavily polluted. Indeed, some earthly creatures can scarcely live—eat, drink, breathe, or sleep. Kari seems to be radiating waves of pain from its many denizens. Alongside the dystopic visions of doomsday, the saviors and utopic callers herald the possibility of new beginnings through radical change of life on the planet. Will the planetary inhabitants figure out the secret of viable life, or will they destroy the planet? Those in Avatara Lokam who believed that the horrors of colonialism, enslavement, genocide, and dehumanization would usher in a future of remorse, humility, cooperation, and benevolence have been thoroughly discredited. Instead, the elite of the planet have grown more daring, more rapacious, more ravenous in their greed.

Epochs ago, the Kankavars, the omniscient architects of galaxies, began experiments on Earth, a planet they nicknamed Kari for its complex combinations. The Kankavars, as their name suggests, are a spectacular species. Technologically sophisticated and deeply empathic, they are capable of infinitely morphing and mutating their capabilities. Their bodies transform to take on diverse morphologies, pigmentation, and sensory capabilities. Their enhanced sensoria are endlessly attuned and receptive to varied stimuli in the galaxies—to new visual, olfactory, sonic, lingual, haptic, hypnotic, gustatory, alimental, and phonic clues. Kari has been to date their most ambitious and beloved project. Here, they are attempting to create beings in their own image, acutely sensitive, infinitely inventive, eternally playful, and continually striving for reflexivity and justice. Since they began the experiment, many iterations have been initiated and failed. In each turn, the Kankavars have learned the unique secrets of carbon-based life and its evolution. In their inimitable fashion, they have gently directed life on earth. For this experiment, they have invented avatars, each endowed with unique capabilities and bestowed with particular responsibilities for evolution of life on Kari. The grand avatars and their majestic personae follow Kari with care, precision, and responsibility. After all, it is their vision and intelligence that help craft the evolutionary possibilities on the planet.

Currently, the experimental iteration #1728 is unfolding on Kari. After beginning their new iteration, tens of millions years ago in current earthly time, and planting secrets in the geological, biological, and atmospheric layers of the planet, the avatars of the Kankavars restarted the theater of life on Kari. However, thousands of generations later, trouble seems afoot. The avatars are furiously consulting each other. It is increasingly clear that Kari is on a path of reckoning. Will the Kankavars be compelled to initiate a new iteration?

An Avatara Lokam, or Land of Lost Dreams, lies beyond the strato-sphere, nestled in the rarified outer folds of the integuments of each "ele-mental" planet. The architects seeded each planet with a different primary set of chemical elements; on Kari it was a carbon-based evolutionary scheme. Each Avatara Lokam is command central for the Kankavars' experiments with elements. As the flows of history meander through the ecstatic and horrific, through war and peace, through the tranquil and turbulent, the lost dreams of generations ride the grooves of billowing clouds to their new abode up in the skies. When souls are lost, relin-quished to the invisible, forgotten in the annals of history or in the memo-ries of those who live on, these souls float up to their new abode above the clouds. In an Avatara Lokam, there is no distinction between the material and immaterial, body and soul, deeds and thoughts. All things on their way to oblivion become liminal entities and ascend to their supraplanetary abode. Forgotten or discarded thoughts, theories, ideas, memories, and knowledges find new homes in the outer reaches of the planet. In an Avatara Lokam they are all transformed into souls.

The lost denizens wander the skies. From time to time, these lost souls are rediscovered, summoned back to Earth as they find a new lease on life in new bodies or to newly found acclaim and cultural revivals. Some descend in new forms or avatars; resurrected souls once again inhabit worldly lives to exult in their recent stardom, or suffer a dreaded infamy. Some rejoice, but others descend unwillingly to face the music of a new generation of earthly theater. The rest explore their new abode, perched among the nooks and crannies between the planets, stars, and clouds to follow the dreams and nightmares of earthly dramas. For some, this is heaven—the immortal television of the afterlife, a device with infinite channels. In their everyday lives, earthlings have free will, and events unfold through the worlds they create. Periodically, the inhabitants discover the "secrets" embedded in their world and learn to enjoy the imaginative potential that the Kankavars had introduced. It was only occasionally, in moments earthlings experience as the "unexpected," that the Kankavars intervened to recalibrate or reroute their experiments.

Of late, the Kankavars are watching Earth with particular care. The experiments on Kari were undertaken to explore whether carbon-based biological evolution of life might engender complex ways of life and ways of collective living. The Kankavars were interested in producing life that preserved harmonious coexistence, alongside a passion for dynamism, diversity, and change. Driven by an ethics of play, justice, joy, and invention, the Kankavars were fascinated with cycles and cyclical life. On Kari, they built a recursive world—one that repeated, restored, recycled, reincarnated, re-engineered, and reanimated life continually in cycles of birth and death, creation and destruction. During each cycle, the planet's Avatara Lokam serves as a holding place for what earthlings reject or denigrate. Perhaps life on Kari will evolve to appreciate the founders' wisdom? In successive renderings, these lost souls are returned to Kari for new iterations of the grand experiment. With each cycle, the Kankavars add new innovations that improve on their growing understanding of carbon-based organic evolution. For the omniscient Kankavars who can see souls of the material and immaterial migrate between worlds, there is a teeming flow of recycled life circulating between Kari and its Avatara Lokam.

On the billowy hills of Earth's Avatara Lokam, one can see the Kankavars deep in thought as they absorb the news from the planet below. Millennia ago when they sowed the seeds of life on Kari, and subsequently through further interventions, the Kankavars have nurtured their experiment. The critical principles of life are enshrined in the avatars assigned tasks, each around a particular node, or Kanu. Each Kanu was tasked with playful intervention at the behest of the founders' vision. The avatars realize that the planet is moving to a denouement. Due to the vagaries of carbon-based evolution, the experiments have not unfolded as they expected. The planet is quickly reaching the point where the lost avatars might need to be awakened, avatars that have been long embedded in earthly biologies, monitoring and overseeing the planet. Once they are unleashed, new visionary tales will emerge as the lost dreams come to reclaim and reseed the imaginative landscapes of Kari. Here are some of their tales.

HOLY SCIENCE

Avatars for Bionationalism

Tales from (An)Other Enlightenment

Perhaps sorcery is necessary to level with this land, to live and walk in the bush of ghosts.

ANNA BADKHEN, "Magical Thinking in the Sahel"

This is a story from our past, from a time so remote that we argue, sometimes, about whether we should call it history or mythology. Some of us call it a fairy tale. But on this we agree: that to tell a story about the past is to tell a story about the present. . . . This is the question we ask ourselves as we explore and narrate our history: how did we get here from there?

SALMAN RUSHDIE, *Two Years Eight Months and Twenty-Eight Nights*

IN INDIA, ONE IS BORN INTO A NARRATIVE WORLD. STORIES and storytelling suffuse life. As children, we lived our days and nights enraptured by stories from parents, grandparents, aunts, uncles, and neighbors. For my generation, once literate, the wildly popular world of the comic book series *Amar Chitra Katha* consumed us.[1] Unbeknownst to our young minds, the series brought themes of Hindu nationalism and conservative cultural politics of gender and caste into a series of now classic characters drawn from mythology (McLain 2009, 2011; Sreenivas

2010).[2] The powerful stories with their vibrant graphics reverberate and endure. Little did we realize that these legendary stories linked masculinity, Hinduism, (fair) skin color, and (upper) caste with ideals of the good and desirable—authority, excellence, and virtue (Amin 2017). These stories successfully circulated Indian mythological imaginations to many multireligious and secular childhoods.[3] For many of us who grew up going to Catholic and Christian schools, multiple religious stories flowed seamlessly. The stories, largely from the Hindu mythologies of the *Ramayana* and the *Mahabharata*, held a cornucopia of modes of storytelling—stories within stories within stories. They helped us discover astonishing new worlds peopled by fabulous characters entangled in the mundane and the fantastical. These stories were rich in intrigue and drama and promised endless possibilities. The characters spanned biologies—animals, plants, humans, human-animal hybrids, imaginary creatures, and fantastical beasts. Monkeys flew across the skies. Humans transmuted, appearing with elephant or lion heads or effortlessly transforming into animal forms. Worlds of humans, gods, and demons coexisted, and individuals moved between these worlds through acts and deeds. The West may have virgin birth, but India has promiscuous possibilities!

The ethos of enchantment and the magical was reinforced with tales of Christian saints, Muslim pirs, local swamis and gurus, spirit figures, jinns, and sacred healing shrines. The promise of good health and wealth always sent people in pilgrimage to temples, mosques, churches, synagogues, and various animist shrines. Tales of human miracles were extolled in stories about saints like Shirdi Sai Baba, Satya Sai Baba, Syed Shah Kareemullah Hussaini Qadri, Kat Bava, Marttaśmūni, Katamattattachan, Sister Alphonsa, and Mother Teresa, among others.

Stories were always multiple, and locally inflected. If I sound too romantic about my childhood, let me temper it. As I elaborate in the Notes to the Mythopoeia at the end of the book, these tales were neither innocent nor apolitical. The stories resolutely secured social hierarchies. And people resisted. Dark-skinned demon gods from North India were transformed by Southern Indians into erudite and revered gods. The patriarchal versions of the *Ramayana* have spawned queer *Ramayanas* and feminist *Sitayanas*. Local animist traditions circulated stories of regional and local spirits, gods and goddesses not represented in the repertoire of upper-caste mythologies. As these stories traveled across and outside India, new variations emerged. The natural, preternatural, and supernatural cohered, as spectral and ethereal worlds mingled with the mythological and historical stories in school textbooks, into a sumptuous blend of this world and others. Worlds commingled—historical figures alongside phantasms and ghosts, gods and devils, demons and angels, and haunting ethereal, incorporeal beings. Stories eschewed the hubris of time, form, corporeality, modernity, science, religion, or rationality. These were the delicious and lavish imaginations of an Indian upbringing. As Vikram Chandra (2014: 42) explains:

> This multiply layered narrative was how I lived within myself, how I knew myself, how I spoke to myself. There was the modern me, and also certain other simultaneous selves who lived on alongside. These "shadow selves" . . . responded passionately and instantly to epic tropes, whether in the *Mahabharata* or in Hindi films; believed implicitly and stubbornly in reincarnation despite a devotion to Enlightenment positivism; insisted on regarding matter and consciousness as one; and experienced the world and oneself as the habitations of *devatas*, "deities" who simultaneously represent inner realities and cosmic principles.

LORD GANESHA'S PLASTIC SURGEON: THE BIOPOLITICS OF HINDU NATIONALISM

The exuberant mythological stories of India have commingled with notions of modernity and orthodoxy, the sacred and the profane, of the centuries-long evolution of science and religion, mutating, shifting, and transforming through the local and the global, and the histories of colonialism, independence, postcolonialism, globalization, and nationalisms. Imagine my surprise when these worlds came hurtling into focus one day.

On October 25, 2014, India's new prime minister, Narendra Modi, while inaugurating a new hospital run by the industrial group Reliance in Mumbai, proclaimed (Rahman 2014):

We can feel proud of what our country has achieved in medical science at one point of time. . . . We all read about Karna in the Mahabharata. If we think a little more, we realize that the Mahabharata says Karna was not born from his mother's womb. This means that genetic science was present at that time. That is why Karna could be born outside his mother's womb.

We worship Lord Ganesha. There must have been some plastic surgeon at that time who got an elephant's head on the body of a human being and began the practice of plastic surgery.

Here the narratives of Indian mythology mingle with the powers of scientific reason to imbue the fantasies of Indian mythology and storytelling with rational possibilities and an energetic Indian scientific prehistory. Perhaps Modi was being humorous or ironic (as some of his supporters have claimed to me). But Ganesha is an elephant-headed god; he is divine. Modi could well have claimed that a god could, through divine powers, connect the bodies of an elephant and human, or that a god does not need circulating blood or a central nervous system! Rather, Modi invented a plastic surgeon to perform an operation to connect the two interspecies body parts. This is precisely the imagination of Hindu nationalism that I find fascinating and significant—science and technology and their practitioners mediate mythological and divine worlds. Even gods need doctors! What we witnessed was the prime minister of India, recently elected by a robust majority in the world's largest democracy, making bold claims of the scientific basis of an ancient Vedic civilization[4].

Modi was not the first. These claims that modern technology, including surrogacy, plastic surgery, genomics, evolution, atomic physics, air travel, chemistry, architecture, fluid dynamics, geology, botany, and zoology, had its roots in the ancient Puranas[5] and the Vedas have been repeated by members of the Hindu right, including various party members and government officials in power (I. Patel 1984; Nanda 2016). We have also seen attempts to link contemporary scientific knowledge with Indian mythology (Ganeshaiah, Vasudeva, and Shaanker 2009; Naresh, Mukesh, and Vivek 2013; Kalra, Baruah, and Kalra 2016; Padhy 2013). Merely writing off Modi's revision of history, his version of Hinduism, or his interpretation and manipulation of science as "silliness" misses the significance and power of the brand of political Hinduism for which Modi stands. At the heart of this contemporary form of Hindu nationalism, *Hindutva* or "Hinduness," is the imagination of a great and grand Hindu past where science, technology, and philosophy thrived. Unlike other fundamentalisms, Hindu nationalism brings together a melding of science and religion, the ancient and the modern, the past and the present into a powerful brand of nationalism, a vision of India as an "archaic modernity" (Subramaniam 2000). Science and religion have long been potent forces within Indian politics and culture, but in recent times they have been mobilized together for a political Hindu nationalist vision. Hindu dominance, intolerance of, and supremacy over other religions, faiths, and traditions, and hatred and bigotry toward non-Hindus, mark the religious nationalist vision. Rather than characterize Hinduism as ancient, nonmodern, or traditional, the Hindu nationalists have embraced capitalism, Western science, and technology as elements of a modern, Hindu nation.[6]

As an apt slogan on a wall once proclaimed, "Be Modern, Not Western."[7] Indeed, India claims its own modernity, and the much-touted Vedic sciences, math, and technologies undergird claims of a great ancient civilization and its own enlightenment. Yet alongside this vision of a scientific and technological superpower, Hindu nationalists espouse a conservative worldview on gender and caste. The recasting and reframing of women as "authentic," as symbols of the nation, signal the superiority of the East, recruiting women into nation building (Sangari and Vaid 1989). Indeed, during a world tour, Prime Minister Modi proudly declared India to be "the only country in the world where god was conceptualized in the female form" (Baruah 2014). Behind this symbolic female power lie a renewed Hindu masculinity and a redomestication of women. Science and religion have proved to be two powerful tools through which religious nationalists have imagined and engaged with this project of masculinization.

With the world listening, Modi, during his first speech as prime minister to the United Nations, proposed that we create an International Yoga Day, since it "could help people connect better with the planet" and could "help us deal with climate change" (*BBC News* 2014). This bold claim of the healing powers of ancient Yoga is similar to his embrace of *Star Wars* in another speech, this time in Madison Square Garden, in which he told his jubilant diasporic Indian audience, "May the force be with you!" (Ridley

2014). Such unabashed proclamations have grown in tandem with a rise of Hindu nationalism as a thoroughly modern, ambitious, global, and organized political power (Hansen 1999). Modi's declaration about ancient India is not mere silliness but reveals a deep-seated belief in the lost greatness of an ancient India, a greatness that the movement means to reclaim. Science and religion here are not oppositional forces but syncretic collaborators.

Behind the euphoric narrative of India as an emerging world power lies a fascinating but untold story of an evolving relationship between science and religion. As India emerges as a global power, the story of science and religion is important because it is precisely here that India's contestations with modernity emerge. To date, there has been little exploration of the relationships among gender, religion, and science. India's narrative of national development and scientific and technological prowess as well as its claims to an ancient scientific and cultural civilization are all key to the country's national identity as an emerging global superpower. This book fills this gap by exploring the enduring relationship of science and religion in India as both forged practices and ideologies that resist gender and caste transformations.

We are witnessing in India, like in many parts of the world, a hypernationalism: nationalistic fervor on multiple fronts. Through five illustrative cases involving biological claims, I explore an emerging bionationalism. The cases are varied, spanning stories of a revival of an ancient Hindu science of architecture, the politics of "unnatural" sex as enshrined in Section 377 of the Indian Penal Code (which the Indian Supreme Court recently ruled on), the veracity of claims of a flying monkey god Hanuman and a bridge he built, debates on the peopling of India through new genomic evidence, the revival of traditional systems of Indian medicine through genomics and pharmaceuticals, the growth and subsequent ban of gestational surrogacy, and the rise of old Vedic gestational sciences. I am fiercely critical of the current configuration of political Hinduism that dominates India, but rather than reject India or Hinduism, I want to embrace them in their more progressive and imaginative possibilities.

It is a blessing to be writing in the twenty-first century, when so much excellent scholarship has already complicated any notion of neat binaries, facile nativisms, innocent nationalisms, or simplistic civilization logics and offered in its place a world teeming with complex global circulations of power, knowledges, resources, and peoples. The field of feminist studies has impressed upon me that there is no sex, gender, or sexuality that is not co-constituted with race, class, caste, and nation. The field of religious studies has taught me that religions and the religious have porous boundaries, bleeding into everyday life, culture, history, and politics. The field of science and technology studies (STS) reminds me that science and scientific knowledge are co-constructed with the social and embedded in its historical and political contexts. Histories of colonialism, casteism, slavery, and patriarchy shape them all. These rich traditions of analyses inform, shape, and animate this book.

In exploring the contours of modernity in India and the confluence of science and religion, one must understand India as a site of a scientized religion and a religionized science (Subramaniam 2000). Hindu nationalists celebrate both the Vedic sciences and the scientific Vedas. Any interdisciplinary exploration necessitates introducing fields to one another. I have crafted this introduction very deliberately into seven sections so each reader can choose relevant parts. Section 1, "Bionationalism," begins with a discussion of why the biological, particularly the biopolitical, proves to be a useful lens for India; it allows us to understand how old, and often conservative, cultural categories of gender, caste, and sexuality are transformed anew in modern scientific discourse. Section 2, "Archaic Modernities," develops the concept of archaic modernities and its gendered dimensions. Biological science, rather than a universal enterprise, morphs into locally modulated sciences. In section 3, "India's Modern Temples," I draw from the history of science and STS to argue for a longer history of the imbrications of science and religion. Science in India has been shaped by its long-enduring relationship with Indian religious practices and cultures. In section 4, "Secularism's Religion," I explain the multiple valences of the term *religion* and define it at its most capacious. I explore the particular relationship of religion and secularism, and how religion spills over into the social, political, and cultural. Section 5, "Lost in Translation," traces the difficulties of writing about the "West and the rest." A Eurocentric focus plagues all academic fields, rendering the rest of the world as "special cases." In this section, I explore why we must, and how we might, decenter the "West" to develop more robust theoretical engagements with the world. I follow with section 6, "Bionarratives," a methodological section exploring the challenges of writing about religion, science, and postcolonialism. I offer bionarratives and helical (spiraling) thigmotropic storytelling as innovations that help us better capture the global travels of science. I end with section 7, "Avatars for Stolen Dreams," a theory of avatars as a conceptual tool for postcolonial biopolitics.

Throughout the book, my aim is to make a sustained and impassioned case for the importance of feminist postcolonial STS, so that the ideas are not relegated to the margins as "just" a particular analytic for India, or "just" for postcolonial nations, but as a central analytic project for all feminism(s), science(s), and STS. In my view, we simply cannot afford not to.

1. BIONATIONALISM: THE BIOPOLITICS OF HINDU NATIONALISM

One of the hallmarks of Hindu nationalism is the centrality of "biology" and the scientific within the imagination, teachings, and practices of a political Hindu nationalism—claims of common blood, indigenous DNA, unique theories, native ecologies, and regimes of bodily discipline are all grounded in a vision of Hinduism as a modern, scientific religion. It is thus a thoroughly scientific nationalism. Since I focus on the biological sciences, I have adopted the word *bionationalism*. First conceived by Gottweis and Kim (2009) in the context of World War II South Korea, *bionationalism* captures the transformation of traditional ethnic nationalism, primarily of blood and group affiliation, into a biopolitical construct grounded in the biological and scientific. I borrow their term to elaborate a bionationalism in the Indian context, where Hindu nationalist ideas and ideologies are scientized through biopolitical claims about gender, race, caste, and sexuality. Essentialist claims of caste continue to be bolstered in a country where caste remains a defining factor. Indeed, no aspect of life in India is untouched by caste—religious, social, economic, and political.

In attempting to trace the imbricated histories of science and Hinduism, I focus on the biological sciences for several reasons. First, Hindu nationalists unequivocally claim the sciences as part of Hinduism and its knowledge system. Biology is central to the imaginations of this thoroughly scientific nationalism. Second, the "biological"—the body and the workings of the body—is central to the philosophies and daily practices of Hindu nationalism. The idea of a militaristic and armed masculinity is deeply rooted in patriarchal conceptions of Hinduism and a biopolitical conception of the body (McDonald 1999). Indeed, bodily discipline through a daily exercise regime, meditation, and yoga is extolled because strong minds need virile bodies. These practices form the cornerstone of training young cadres of Hindu nationalist grassroots organizations such as the Rashtriya Swayamsevak Sangh (RSS) in neighborhood *shakhas* and schools all across India. Physical discipline is seen as an essential element for overall discipline; a strong body produces a pure mind (Chitkara 2004). However, women can embody such nationalist honor only if they remain chaste and virtuous, never challenging patriarchal Hindu family values (S. Banerjee 2006). Third, biological metaphors and concepts litter the landscapes of the Hindu "body politic"—from metaphors of "pure" Hindu blood, blood ties, tissues, and DNA to metaphors of biological geographies, landscapes, and ecologies. The alleged hyperfertility of minority women, especially Muslim women, continues to haunt Hindu nationalist anxieties. The liveliness of biology is important to the vitality of politics and social movements. Fourth, the biological transcends Hindu nationalism. Deeply imbricated in religion and science, the biological sciences are central to conceptions of the body, of health and illness. In India, and its cultural imaginations, the boundaries between body and spirit, human and nonhuman, living and nonliving, and natural, preternatural, and supernatural have long been incredibly porous. Global universal histories miss the particularities of local contexts and transformations as well as the local and daily practices and processes (Phalkey 2013).

Fifth, tracing biological knowledge allows us to see the converse—how local biological knowledge and categories are deeply connected to global preoccupations. Biological claims of gender, race, caste, and sexuality have been a mainstay of scientific knowledge (W. Anderson 2002a; Philip 2004). We can trace how science's oppressive histories (for example of race and caste) find new reinstantiations. Old stories are retold in new scientific language, the new narratives quickly adapting to shifting power structures in India. Striking is its recursive nature—certain political concerns return to haunt us repeatedly. Finally, recent innovations in the history of science propel us to move from seeing science as purely an ideological tool of empire or the state (Nandy 1988; Baber 1996) to seeing it as a set of model knowledge systems that emerge from particular historical, economic, and political contexts (Ganeri 2013). Focusing on biological knowledge in India allows us to trace local practices and demonstrate how religion and science can come to be deeply and inextricably interconnected.

I draw on Foucault's notion of biopolitics, recognizing that it emerges from a Western colonialist genealogy, and that biopolitics in the colonial and postcolonial worlds need to be understood differently (Legg 2007; Stoler 2002). While Foucault introduced the terms *biopolitics* and *biopower*, he did not fully describe or explain them in his works and often used them in conflicting and even contradictory ways. But the concept captures something important, as evidenced by the "intensity of the debate and the prominence of biopolitics" (Lemke 2011), and it remains a useful concept to think and work with, even in contexts for which it was not developed. In a singular focus on tracing the circuits of power in the West, Foucault theorizes a significant shift in power in the eighteenth century from "sovereign power" into state "biopower." He argues that societies witnessed a transformation from the old "right to *take life* or *let live*" to a new "power to *make live* and *let die*" (emphasis mine). In short, Foucault ([1976] 1998: 137) described biopolitics as life, and its governing imperative "as a form of power that exerts a positive influence on life, that endeavors to administer, optimize, and multiply it, subjecting it to precise controls and comprehensive regulations." Biopower is not top-down and imposed by a sovereign power but diffuse and embodied in discourse, knowledge, and "regimes of truth." As Foucault shows, biopolitics transformed the effects of power from a negative (something that excludes, represses, masks, and conceals) to a positive (something that produces). By the nineteenth century, biopower's dual focus on the "anatomo-politics of the human body" and "regulatory controls" of population came together through a series of "great technologies of power." We witnessed new kinds of political struggle "in which 'life as a political object' was turned back against the controls exercised over it, in the name of claims to a 'right' to life, to one's body, to health, to the satisfaction of one's needs" (Rabinow and Rose 2006: 196). Rather than be forced or coerced into controlling and managing our biologies, we voluntarily submit to regimes of control. For example, eugenic logics today are not always enacted through "forced sterilization" but often in regimes under which individuals are said to "volunteer" to terminate pregnancies out of free will and "choice" (D. Roberts 2009).

India reminds us that the history of biopolitics is uneven, and biopolitics in India needs to be understood in a different register from the West, more as a layered history of the precolonial, colonial, and postcolonial. It is in part shaped by Western practices of governmentality, institutionalized through the British Empire during colonial rule but also through biopolitical practices deeply embedded in local practices and religions. While this book largely focuses on bionationalism in recent years and its striking characteristics with the rise of Hindu nationalism, as you will see, bionationalism has a longer history through earlier religious nationalist and secular governments. In particular, politics of gender, class, caste, sexuality, and nation have long shaped the biopolitical contours of India, albeit unevenly and heterogeneously. Although varied geographically, strict laws of purity and pollution, regimes of body care, and codes of endogamy have long governed South Asia. British colonial rule brought with it the practices of Western governmentality and its biopolitical dimensions that came to confront local biopolitical practices in the subcontinent.[8] In this sense, we can understand colonial rule and the resistance to colonialism as a clash of competing biopolitical claims (Prakash 1999). There was no easy alliance, and anticolonial struggles and the independence movement at times contested and at times colluded with the British, helping shape a new biopolitical order that has in turn strongly shaped postcolonial biopolitical governance.[9] This postcolonial biopolitics has given us new formations of contemporary biopolitics in India, the central preoccupation of this book.

2. ARCHAIC MODERNITIES: SCIENCE AND RELIGION IN CONTEMPORARY INDIA

In one of the many memorable lines from *Midnight's Children*, Salman Rushdie (1981) argues that Indians have a poor sense of time and history, "No people whose word for 'yesterday' is the same as their word for 'tomorrow' can be said to have a firm grip on the time." While true of some but not all Indian languages (for example, *kal* means "yesterday" and "tomorrow" in Hindi), this statement captures something important. Indians who visit the United States find the culture's relationships to time and history strikingly different from their own. In its most obvious manifestation, in India we routinely see structures that are hundreds of years old, uncared for, and withering away in sight. In contrast, structures just decades old in the United States are quickly memorialized, often with an entrance fee! To understand the Indian past is to enter time warps in which the "silent and evasive" pasts come to the fore in contemporary India (Nandy 2001; Subramaniam 2017). Time-folding warps are one of the critical ingredients of Hindu nationalism. Hindu nationalists bring the past and present together into one seamless story of a past sutured to the present, with a firm excision of the middle years of colonialism and conquest, in particular the histories of Islam. The historical record reveals a vastly more complex past. Repeatedly in the cases presented in this book, temporalities are critical—shifts where the ancient and modern cohere. However, it is not only Indians who can live in multiple times. While it is rarely acknowledged to be the case, Europe does too (Chakrabarty 2001; Bauman 2013).

India's archaic modernities and the articulations of the new *Hindutva* are transforming India's foreign and domestic policies, as well as public and private lives. The coming together of science and religion is not only a personal vision but also one that brings configurations of science, religion, and corporate and political power. For example, Baba Ramdev, a swami in saffron robes and a flowing beard, illustrates the archaic modernity I am trying to describe. He claims to be "scientific, secular, and universal," yet he also claims that he can "cure" homosexuality with yoga and has openly fantasized about beheading those who refuse to chant nationalist slogans (Crair 2018). A *sanyasi*, pledged to spend the rest of his life as a Hindu ascetic, he is also a yoga megastar on national TV, yet he has also been accused of violence, including murder. Nonetheless, he controls and is the public face of Patanjali Ayurved Ltd., a multibillion-dollar corporation that celebrates ancient Vedic heritage through consumer products including toothpaste, hand soap, floor cleaners, toilet cleaners, and more recently instant noodles, organic grains, pulses, and herbs. Baba Ramdev argues that he runs the company not as a CEO runs a business but as a "guru runs an ashram." Employees are forbidden to eat meat or drink alcohol, and their labor is a form of *sewa*, or spiritual service, commanding lower salaries as a result (Crair 2018). There is now an explosion of consumer products based on Ayurvedic principles. Even foreign multinationals have created Ayurvedic consumer products: Colgate's Vedshakti toothpaste, Lever's Ayush toothpaste, Tresemme's Botanique, and Clinic Plus's Ayurveda (Sachitanand 2017). He effectively uses the mantle of *swadeshi* (or Indian economic nationalism) to reframe foreign companies as neocolonial villains (Worth 2018), effectively creating a "state-temple-corporate complex" (Nanda 2011). This quintessential blending of science, religion, and capitalism in an archaic modernity is in view throughout contemporary India.

To be sure, I do not want to suggest an abrupt and sudden arrival of an archaic modernity. Rather, the idea itself can be said to have multiple temporalities. Indeed, throughout India's history, in multiple versions and visions, the past and future have cohered. Many Indian thinkers and leaders have rightfully celebrated and drawn from South Asia's significant historical, cultural, and philosophical heritage. At independence, in refusing to call the country Hindustan and claiming for itself the older name of India or Bharat, the nation presented itself as an inevitable, organic outcome of national movements since the late nineteenth century. Thus, even while Pakistan emerged as a "new" nation, "India seemed young only in age, but ancient in spirit" (Chakrabarty 2018). As we see in chapter 1, the Indian nationalist movement, in attempting to blunt the claims of colonialism, drew on India's ancient history to claim a superior civilization. Nehru (1959), despite being staunchly secular, multicultural, and a strong supporter of science and technology, tempered his passionate embrace of science with a celebration of the wisdom of ancient India. In his *Discovery of India*, he waxed eloquent about the unique contributions, rich heritage, and philosophical and cultural significance of ancient India (Nehru 1989). In contrast, Marxist historian of science Debiprasad Chattopadhyaya brought past and present together in a different way (R. Prasad 1993). Rather than characterize the past as idyllic and harmonious, Chattopadhyaya emphasized class struggle and the vibrancy, debates, and clash of ideas that animated Hindu philosophy.[10] Thus, the temporalities of India have always defied a linear historical narrative.

There have been profound political and economic shifts in India since the increasing liberalization and globalization of its markets starting in the 1990s. The imbrications of nation, science, and religion have largely worked together to respond to this globalized phase of capital. It is important for us to understand the deeply entrenched logics of *Hindutva* beyond the politics of the current government or the BJP. I hope that the book, as it unfolds, presents *Hindutva* as a deeply embedded ideology, sometimes unnamed and at other times (like the contemporary moment) overdetermined, but always entangled with other forces such as capital, Eurocentrism, orientalism, postcolonialism, "third-world" politics, pressures from global financial institutions, and trade laws. The proliferation of Hindu swamis and gurus, and their capitalist and consumerist infrastructures, help elevate and domesticate nationalism into the folds of everyday life and a "biomoral consumerism" (Khalikova 2017).

3. INDIA'S MODERN TEMPLES: THE INTIMATE HISTORIES OF RELIGION AND SCIENCE

Soon after independence, Jawaharlal Nehru, India's first prime minister, famously declared India's massive hydroelectric dams and gleaming new national scientific laboratories the "modern temples" of India, built in "service of our motherland" (D'Souza 2008; Arnold 2013). Fifty years later, Pope John Paul II, during his visit to India, sat down with the leaders of ten other religions. All the leaders were men, and they met in the capital's Hall of Science. Seventy years after independence, a Hindu nationalist government emerged, running on a platform of "development nationalism." These three moments serve as important signposts for postcolonial India. In each, we see the visible entanglements of science and Hinduism. How should we understand the very visible and public encounters of two institutions—science and religion—in India?

While Hindu nationalists are reinvigorating a vision of archaic modernities in contemporary India, the intertwined histories of science and religion run deeper. As Gyan Prakash (1999) argues, "Standing as a metaphor for the triumph of universal reason over enchanting myths, science appears pivotal in the imagination and institution of India, a defining part of its history as a British colony and its emergence as an independent nation." Indeed, during India's freedom struggle, the nationalists opposed British colonialism in the name of "reason" and claimed "reason" as an Indian invention, not a Western one (M. Chatterjee 2000). They claimed that India's ancient history had "followed, if not pioneered, a universal spirit leading to the nation-state, republicanism, economic development, and nationalism that reaffirmed cunning of Reason; and [they asserted] that a 'backward' country like India could modernize itself, if liberated from colonial slavery" (Prakash 1990: 391). Unlike linear diffusion models (Basalla 1967) that imagined colonized countries as passive grounds for the imposition of a "superior" Western science and civilizational logics, India (and indeed all colonized countries) clashed and resisted (Raina 1996, 1999, 2003; Raj 2001; Philip 2004; Chakrabarti 2004; Raina and Habib 2006; Prasad 2008; Chambers and Gillespie 2000; Loomba 2015). Rather, science in India progressed through "negotiations" and a process of "hybridization" rather than any wholesale imposition by the West.

Historians of colonial science have, by now, amply demonstrated that we can no longer assert a unique "Western" genealogy for modern technosciences. Instead, technosciences were already global by the nineteenth century (Teresi 2001; Mukharji 2016). Colonial science in India tugged the technosciences away from their Western roots and combined it with Indian, specifically Hindu, forms of knowledge (Prakash 1990). This evolving new science was not simply derivative of the West. It "drew its authority from a reworking of India's 'archaic' traditions and a re-siting of these within the matrix of the modern nation. India's modernity, with science at its core, was thus at one and the same time both Indian and Western" (Arnold 2000: 162). My book is inspired by the work of South Asian scholars who have documented the rich local and vernacular cosmologies of South Asia (Arnold 1993, 1996; Philip 2004; Aditya Bharadwaj 2016; Sur 2011; Philip, Irani, and Dourish 2012; Abraham 2013; Phalkey 2013; S. Pinto 2014; A. Prasad 2014; Mukharji 2016). Ambivalence and hybridity are the hallmark of such engagements.

While the colonized resisted any easy imposition of science by colonial powers, colonial powers have historically controlled the narratives of science and its biopolitical imperatives. Colonial policies were fundamentally biopolitical in nature, producing biological "difference" and keeping those differences in their "proper" places (Adas 1989; Schiebinger 1989, 1993; Stepan 1991; W. Anderson 1995; Sylvester 2006). In countries like India where the colonial enterprise depended heavily on the involvement of natives for governance, colonizers simultaneously maintained that the natives were worthy objects for their civilizing mission yet could never be actually civilized completely (Prakash 1999). Natives were always seen as "fixed" and static while the European character was never fixed and was open to growth and change (Stoler 2002). Theories of biological superiority and inferiority produced ideologies of difference—of sex, race, caste, sexuality, class, nation—determining which bodies were capable of being subjugated (Hammonds 1999; Harding 1993; Stepan 1982). Claims of "irrational" natives, with "hot" temperaments and promiscuous sexualities, bolstered logics of difference and thus colonial subjugation. The deep misogyny that shaped male colonists is reproduced in policies of citizenship, governance, and public and private rights, and ultimately of the embodied subjectivities of colonizers and colonized (W. Anderson 2002b, 2007; Stepan 1982; Harding 2008, 2012). Deeply entrenched in theories of the natural sciences, the legacies of these colonial ideologies of difference continue to haunt contemporary biological theories of difference (Subramaniam 2014). In sum, we should understand the sciences as "sciences of empire" (Schiebinger 2004) or as "Northern ethnosciences" (Harding 1997) that have fundamentally developed in a "colonial context" (Seth 2009; Philip 2004). In the colonies, the colonizers' logic of difference merged with local hierarchies of difference to create new biopolitical regimes. For example, in the Indian context, theories of racial differences confronted hierarchies of caste to produce new theories and scientific analogies between race and caste.

Science and the Postcolonial

Given that science and technology have been "the jewels in the crown of modernity" (Harding 2012: 2), central to the expansion of empire and critical to the contemporary world (Schiebinger 2005; Schiebinger and Swan 2005; Harding 2008; Subramaniam 2015b; Foster 2017), how do these colonial ideologies shape postcolonial India? There is little agreement among scholars on what we mean by *postcolonial science* (Abraham 2013) or, for that matter, by *colonial science* (Phalkey 2013). While not a seamless continuation of colonial science, postcolonial science may be understood as science "rephrased within the framework of globalization," making visible colonial ontologies that live on in other guises and names (Anderson and Adams 2007). Postcolonial studies of science offer us "flexible and contingent frameworks" to better understand the travels and multiple "contact zones" of technoscience and empire across laboratories, cultures, societies, and nations, where these encounters across difference have engendered new cultural forms (Pratt 1992; Verran 2001, 2002; W. Anderson 2009).

A critical reason for colonialism's immense and enduring power is that the elites of postcolonial states embraced science and technology as a central mode of modernity and development (Abraham 1998, 2006; Droney 2014; S. Roy 2007). While colonial and postcolonial STS have in recent years challenged and renarrated the history of science from earlier hagiographies of the superiority of Western science and its linear diffusion to the colonies, the old colonial hagiographies still predominate in India's own elite histories of science (Prasad 2014). This book, in the tradition of recent work in postcolonial STS, seeks to displace these older imaginings of Western supremacy with new accounts that emphasize the resistance, hybridity, and contested nature of science in India (STHV 2016). Yet colonial policies do endure in the civilizing logics of contemporary international development projects (W. Anderson 2009; Escobar 1995), and contemporary scientific theories and practices continue to show the deep legacies of colonialism (Harding 2008, 2012; Traweek 1988). Contemporary scientific institutions and education in postcolonial countries often reinforce models of "Western science," replicating Eurocentric models of institutions and knowledge in postcolonial contexts and naturalizing science as "Western science" (Mutua and Swadener 2004; N. Kumar 2009; Thomas 2016, 2017; Krishna and Chadha 2017). The enduring legacies of colonial science in postcolonial India remain deeply fraught and debated.

Early third-world critics of science placed the failure of postcolonial states squarely at the feet of science and technology as a mode of postcolonial national development (Alvares 1992; Nandy 1988; Shiva 1989; Mies and Shiva 1993; Viswanathan 1997). In this more radical critique, critics argued that postcolonial nations embraced science and technology as the "reason of state" (Nandy 1988) and with it embraced the "violence" of Western science (Shiva 1989; Alvares 1992). Arguing that science and technology enabled colonialism, they contend that the third world has been poorly served by modern science and technology, its philosophies, policies, and practices, and are responsible for the economic, political, social, and cultural regress of formerly colonized countries (Harding 2009). These critics contend that postcolonial nations need to retreat from the hegemonic structures of science in order to truly "decolonize" formerly colonized worlds and their knowledge systems.

Decolonization is important, these scholars argue, because it will liberate and develop indigenous, alternate, and local knowledge systems. However, other postcolonial critics argue that this more radical position in many ways reinforces the binaries of center and periphery, West and East. They point to the inherent nativism in these narratives, wherein Western science and technology is always destructive and oppressive while local knowledge systems are always good and liberatory (Nanda 2001, 2003; Phalkey 2013). Postcolonial STS challenges us to move the binary logic to understand the center and periphery as co-constituted, the periphery continually remaking the "center" (Raina 1996; Philip 2004; Abraham 2006). Rather than locate the postcolonial in the third world, they have stressed the circulation of an increasingly transnational science and the creation of new complex formations that resist easy categorization (Fujimura 2000; Raj 2013; Prasad 2014).

Perhaps the best-developed ideas surround recent discussions of "Asia as method," a plan to reorient STS so that Asia isn't just a geographical site for data extraction or a place where European concepts diffuse (K. Chen 2010; W. Anderson 2017). These discussions have pushed STS to rethink science studies in more relational terms—to take into account contested and politicized terrains of history and contested claims of the past. Chen (2010) suggests that "using Asia as an imaginary anchoring point can allow societies in Asia to become one another's reference points" (xv), thus multiplying "frames of reference in our subjectivity and worldview, so that anxiety over the West can be diluted, and productive critical work can move forward" (223). While paying heed to the dangers of narratives framed as a "clash" of knowledges of West and East, and wary of conceptions that bolster essentialized views of "Asian values" (Abraham 2006: 210), scholars have nonetheless continued to build a substantial body of work focusing on science in Asia. Indeed, the frame of bionationalism was first articulated in the Asian context, centered on countries using biomedical technoscience to optimize population health and to aggressively defend the nation against attacks from the outside (Gottweis 2009; Gottweis and Kim 2009). Indeed, the journal *New Genetics and Society* (2009) devoted an entire issue, *Biopolitics in Asia*, to exploring the tremendous resources and development poured into science and technology in China, South Korea, Singapore, Vietnam, and India. The successes of Asian science have not gone unnoticed, engendering calls within STS to develop a postcolonial version of the principle of symmetry so that Asia is not continually understood through the lens of the West and dominated conceptually, linguistically, corporeally, metaphysically, and institutionally by Euro-American thought (Law and Lin 2017). This book responds to these calls.

Tracing the Contours of Bionationalism: Summary of the Case Studies

Rather than reanimate the antagonistic debates over whether science or religion will triumph, I suggest that the critical question is not *whether* the two are related but rather *how*. Over the last three decades, I have tracked the biological sciences in India. Five illustrative cases animate this book, each exploring various dimensions of bionationalism. In chapter 1, "Home and the World: Modern Lives of the Vedic Sciences," I show how bionationalism includes visions of scientific development, best exemplified in the election of Prime Minister Narendra Modi, who ran on a platform of development nationalism. Promising scientific "development," Modi brings together claims of the greatness of the Vedic sciences and promises of a scientific modernity. Using the case of Vaastushastra, or the Vedic "science" of architecture, I examine the valorizing of Vaastu as a modern-day "Vedic science" and its commodification in contemporary India as the best of the ancient and modern, an exemplar of an archaic modernity.

In chapter 2, "Colonial Legacies, Postcolonial Biologies: The Queer Politics of (Un)Natural Sex," I begin with the infamous Section 377 of the Indian Penal Code. Introduced during colonial rule, it remained law until September 6, 2018, when the Supreme Court struck down the sections that criminalized consensual homosexual sex between adults. As we shall see, other acts deemed "unnatural" and "against the order of nature" remain prohibited. But what is nature's order? Scientific conceptions about "nature" and "human nature" shaped the penal code in its colonial formation. I explore how modern Hindu nationalism, Christian and Victorian sexual politics, scientific theories of sex and sexuality, and eugenic policies together shape the emerging bionationalism of sexual regulation. Colonial imaginaries through religion and science shape sex and sexuality. They reify "Indian" bodies and sexuality in opposition to the West, even as the West has shifted its cultural registers.

One of the sites that best illustrate the postcolonial predicament is environmental politics. Drawing on the mythology of the *Ramayana*, I explore the nativist politics that haunt postcolonial environmentalism. In the debates surrounding of the building of the proposed Sethusamudram Shipping Canal between India and Sri Lanka, two sets of actors opposed the construction—environmental scientists and religious activists, albeit for different reasons. In chapter 3, "Return of the Native: Nation, Nature, and the Politics of Postcolonial Environmentalism," I explore the debates over the proposed canal. A central point of contention focused on the underground rock formation at the bottom of the ocean that needed to be destroyed to build the canal. Were these structures natural limestone shoals produced in the last Ice Age, or were they remnants of a bridge built by Hanuman and his monkey army as narrated in the Indian epic *Ramayana*? This case powerfully demonstrates the strong animist traditions of India, where claims of monkey gods and old mythologies emerge as facts to be adjudicated by the courts. In a related case, I explore how demographic histories of caste are essentialized, orientalized, and taken up in the biological literature as a case of ecological sustainability. Both cases together give us a complex picture of the nativism that undergirds postcolonial environmentalism.

Chapter 4, "Biocitizenship in Neoliberal Times: On the Making of the 'Indian' Genome," perhaps best epitomizes bionationalism. Here I explore recent debates on the biological relationships between and among different ethnic and caste groups of India, especially how these relationships relate to theories of an Aryan migration into India. This case demonstrates how old ethnic and caste categories get new life in the language of genomics. Exploring two recent political debates on Indian genomics in India and the United States, I show how old politics of caste and race animate contemporary politics of belonging and are adjudicated through their biopolitical dimension, in the language of genes and genomics. I also explore the emergence of the Indian Genome Variation Project and projects such as Ayurgenomics (aligning Ayurveda with modern genomics), tracing how bionationalism has empowered third-world nations to assert their genomic sovereignty in the face of global genomics. The government has aggressively promoted traditional Indian systems of medicine, including a 2017 bill that, if passed, will allow doctors practicing traditional medicine to prescribe allopathic medicine after a "bridge course."

In chapter 5, "Conceiving a Hindu Nation: (Re)Making the Indian Womb," by tracking the life of the "postcolonial womb," I explore how "women" are figured in visions of an archaic modernity. How did the excesses of the Indian women's undesirable hyperfertility of overpopulation get transformed into the desirable (re)productive site of the womb in neoliberal times? How does bionationalism transform the role of women from the victimized overpopulate, and the modern entrepreneur, to the sacred mother? We see how with the rise of Hindu nationalism, the new entrepreneur once again is transformed to dutiful wife, mother, and daughter of family and nation. In a newer chapter to this history, we are witnessing the rise of ancient Hindu gestational sciences such as *garbh sanskar*. The chapter explores the promotion and proliferation of these practices.

Indian biologists have vigorously contributed to these questions both within national biological research and through transnational alliances. Indeed, much of science today happens through transnational collaborations. Each of the five illustrative cases has led to public euphoria, debates, and contestations in which the ivory tower confronts local, national, and transnational politics. In exploring multiple facets of these five cases, I trace the production of scientific knowledge in and about India. The cases also reveal that the central preoccupations of the nation permeate debates on science and religion. Politics of gender, race, class, caste, sexuality, and indigeneity are deeply implicated in the projects of the nation, and in the alliances and tensions between science and religion. When the encounters of science and religion are examined through postcolonial eyes, science emerges as powerful and with multiple valences, but predictably, it is most effective when its interests ally with other forms of power. The book acknowledges and traces these shared histories to uncover new circulations of science and its role in modernity.

4. SECULARISM'S RELIGION

While *religion* is a ubiquitous term, scholars have a surprisingly difficult time defining it (Nongbri 2013). At its most capacious, "religion is a kind of inner sentiment or personal faith ideally isolated from secular concerns. In this common framework, the individual World Religions are thought of as specific manifestations of the general phenomenon of religion" (Nongbri 2013: 7–8). But the definition ill represents religion's profound entanglements with politics and power. The enlightenment narrative, as Janet Jakobsen and Ann Pellegrini (2008) note, separates secularism from religion, and through such a separation, secularism, like reason, is granted a universal status. In contrast, religion remains particular. Thus, religion emerges as secularism's conjoined twin (Asad 1993). It is secularism that marks modernity's "religion" as a discrete category and institution. Yet, as their insightful work demonstrates, the binary conceptions of secularism and religion are poorly reflected on the ground everywhere in the world. The "resurgence" of religion across the globe reveals tensions around the politics of secularism and its enactments.

In fully exploring the burgeoning Hindu nationalism in contemporary India, it is easy to overstate the case and not adequately capture India's multireligious histories. For example, Sugirtharajah (2018) argues that Indians (and Asians more broadly) embraced the multiplicity of religions and made it their own. In India, Jesus was transformed into a Jain pilgrim and a Hindu mystic. Such transformations were an act of anticolonialism, "an intentional, deliberate, and dignified method of self-discovery and decolonisation in the face of colonial degradation. . . . Their articulations can be seen as a notable attempt at 'provincializing Europe' and a rejection of the notion that only the West can prove the pathways to understanding Jesus. These Asian thinkers demanded a different foundation for faith than history, logic, and neutrality" (Sugirtharajah 2018: 1–2). Sugirtharajah argues that through centuries of engagement, Christianity in India has become "Indian" because the Indians have rigorously engaged its tenets, transforming it and adapting it to Indian sensibilities. We can similarly narrate the other religions in India to reveal how through centuries of practice, each has become deeply rooted in an Indian ethos. In this sense, they are every bit as "Indian" as Hinduism. Religions, like science, always come to develop new, locally inflected roots.

The founders of India constituted a democratic country with a secular and pluralistic vision that was to actively include, support, and encourage all religions (in contrast to an American model of a separation of church and state) (Das Acevedo 2013). The legal system includes exceptions for religious personal law in areas of marriage, divorce, inheritance, adoption, and maintenance. Despite this inclusive vision, there is little doubt that religious intolerance has always persisted. In contemporary India, religious minorities are harassed, intimidated, and sometimes killed in the name of the majority Hindus (Peer 2014; Schultz and Raj 2017). Interreligious and intercaste relationships are often met with violence, and movements such as "love jihad" and "ghar wapsi" are orchestrated campaigns to maintain the social bounds of religion (Faleiro 2014). Even when individuals convert to other religions from Hinduism, their caste origins (and social privileges and stigmas) travel with them.

I use the term *religion* in this more capacious understanding as a social and intellectual institution with multiple genealogies and valences. In particular, I draw on recent theorizations of religion from STS. First, I engage with religion as a set of social practices that have been integrated into mainstream culture and everyday life. Throughout this book, we see how everyday forms of religiosity are intertwined in structures of biodivinity, such as Vaastushastra, an ancient science of architecture that gets incorporated into ideals of the modern home, or how people (including doctors) may pray in their quest for fertility or good health, or how everyday practices of religious life work as principles of conservation and sustainability. Religion in India—or indeed in any part of the world—is rarely set apart from life, politics, or science.

Renny Thomas (2016), in his insightful ethnography of Indian scientists, demonstrates that science does not replace religion in India but lives alongside it in an easy and effortless cohabitation. He argues that one sees the limits of conceptions of Western atheism in capturing the everyday life of scientists. Indian scientists insist on a more engaged relationship with religion because, they argue, it makes them better scientists—for example, by preparing them to deal with complex ideas such as the "infinite" (which is unpredictable, unfathomable, complex, and not measurable) or through mediating "their ego and jealousy" (Thomas 2017). Indeed, scholars have noted such imbrications of science and religion throughout the world. As Elizabeth Roberts (2016: 209) eloquently describes in her work on assisted reproduction in Ecuador, "I came to understand that God was part of the action. Lab biologists included God in the process of assisted reproduction, since 'all of nature was thanks to him.'" And Natasha Myers's (2015) careful ethnography of the life sciences, *Rendering Life Molecular*, reminds us how scientists in the West ultimately willfully fail to "disenchant their practice or deanimate their objects." Indeed, despite the Enlightenment, STS reminds us that the "objective" sciences have not succeeded in entirely dis-enchanting our deeply enchanted world. Despite science's avowed secularism, these works remind us of the enduring Judeo-Christian legacies in the sciences (Noble 1992; Hannam 2009).

Second, I explore religion in its more politicized incarnations when it is mobilized toward larger national and instrumental goals. The heart of the book examines how ideas of the native and indigenous are mobilized into a potent form of bionationalism. Finally, the book draws on contemporary scholars' critiques of the binaries of the religious and the secular. Scholars have underscored that theorization of "religion" in STS rarely embraces the range of social practices, spaces, and beliefs that are connected to sacrality in many non-Western regions, for example, in seeking to recover the differences between words like *dharma, mazhab, karma*, etc. that do not map precisely onto religion.[11] Indeed, this conceptualization of religion is well laid out in Ian Whitmarsh and Elizabeth Roberts's (2016) advocacy for a "nonsecular anthropology." They argue that "secular" science/medicine is predicated and relies on religious traditions to produce science's political secularity. Rather than an areligious institution, Western science emerges from genealogies of Christianity, and Western scientific and medical practices are deeply embedded in Christian ideologies and mores. Indeed, the idea of secularism is itself constructed and shaped by specific Christian traditions (Jakobsen and Pellegrini 2008). As Whitmarsh and Roberts (2016: 203) argue, "Within this analytic, the secular does not serve as a simple contrast, a word for whatever is not religious. Instead the secular is what claims to craft 'the religious' as its object, asserting itself as equally positioned vis-à-vis all religions." *Holy Science* builds on this work to better articulate within STS how the allegedly separate zones of the secular and the religious are coconstituted. Indeed, these disjunctures become all the more apparent in dealing with non-Christian cosmologies. For example, Jean Langford (2016), in analyzing death in the modern hospital, explores the medicalization of the soul. She shows how Western hospitals reveal their latent Christian theological presuppositions when confronted with patients who are adherents of alternate cosmologies about the body and soul, life, and death. Christian theology thus gets encoded in the mask of secular science. Throughout this book, we encounter the limits of Western secular STS to deal with postcolonial and alternate cosmologies and frameworks.

5. LOST IN TRANSLATION: THE ELUSIVE LANGUAGE OF THE COLONIAL AND POSTCOLONIAL

I should state clearly at the outset that in uncovering India, and in exploring the many elisions and erasures in its histories, genealogies, traditions, and practices, I inevitably reinforce the idea of a monolithic "West"—one that is hierarchical and riddled in binary logics, and in which modern, scientific, and abstract thought have deep and enduring roots. As I have circulated this manuscript to colleagues who work in other national, religious, and regional contexts, I have been struck by the numbers of times they have scribbled or commented in the margin, "But this is also true of X, Y, Z." Yes, this is a book about India, but it is written firmly within the shared contexts across the "postcolonial." Similarly, the major religions share characteristics. While the intimate context of this book may be India, its content speaks to these global shared histories.

Without a doubt, the West is also multiple, polyvocal, and nonmodern in its own way (Haraway 1985, 2003; Harding 1991, 1998; Herzig 2005; Latour 1993). Despite this truism and the call to provincialize Europe almost two decades ago, discussions of a monolithic "West" lives on. The spirit of the book responds to how these larger topologies of West and East remain central and enduring civilizational logics that haunt academia. They have shaped not only colonial justifications but also disciplinary practices. For example, in discussing theories of the novel and its form, Amit Chaudhuri (2017) insightfully observes that when greatness is bestowed on a European novelist, it is to credit the novelist with "innovations in the form." In contrast, the Indian novelist is seen as writing only "about India." Innovations of the European novel are seen not as an assertion of Europe but as contributions to the universal form of the novel itself. Formal innovations in a non-European novel may at best be credited with having "a European air." With the growth of American writers and the centrality of the Anglophone world of the USA, it has joined Europe, and so "we don't think of innovations in fiction emerging from these locations as being primarily connected to what it means to be a New Yorker, or an American—we think of them as formal innovations in themselves." Chaudhuri concludes, "The American writer has succeeded the European writer. The rest of us write of where we come from."

Postcolonial scholars have made the same argument from multiple disciplinary locations. It is true of the field of history, where "Europe remains the sovereign theoretical subject of all histories, including the ones we call 'Indian,' 'Chinese,' 'Kenyan,' and so on . . . all these other histories tend to become variations on a master narrative that could be called 'the history of Europe'. In this sense 'Indian' history itself is in a position of subalternity; one can only articulate subaltern positions in the name of this history" (Chakrabarty 1992).

Anthropology has long constituted the European center through the demarcation of the "primitive" other. Sociology, political science, economics, and virtually all disciplines constitute their fields through universal theories of the Enlightenment and with Europe and the West at their center. As historians of science have shown, we see this pattern literally written on the body by the machineries of science and technology. In the biological sciences, hierarchies of bodies and their capabilities were written into conceptions of an evolutionary ladder or a "great chain of being." The "higher races" were capable of rational abstract thought, while lower races were more animal-like, more of the earth, closer to nature (Schiebinger 1993). Women also ranked below men in these capacities (Schiebinger 1989). Thus, third-world peoples and races, and similarly women, are always seen as closer to nature, more grounded in the material body, and its sensory and affective apparatus, and less in the mind (Hamilton, Subramaniam, and Willey 2017). The normative body was always the elite European, and the white European heterosexual couple represented the pinnacle of evolution (Markowitz 2001). Only European bodies marked superior beings, and only European science made the rules.

Similarly, modernity belongs to the West. If others showed some aspects of it, they may have "alternate" modernities or display "multiple" modernities, or in my invention here "archaic" modernities, but the idea of modernity as inherently Western remains canonical. These "rules" of knowledge were not lost on the colonized. Colonized subjects often claimed their own modernity. We see claims to Islamic modernism, Hindu modernism, and so on. But as Lila Abu-Lughod (1998: 128) writes, "these alternate versions were often written off as failed imitations of the west—failures of secularism, of nationalism, or of enlightened modernities. From such language of failures, we need to move to expand our repertoire of modernities, to capture more of the many entangled histories of west and the rest." This repertoire must also capture the "the ambivalences, contradictions, use of force and the tragedies and the ironies that attend to it" (Chakrabarty 1992). Understanding Indian bionationalism is one such attempt.

If the colonized and the postcolonized are forever to write in the language of the colonizer, is much lost in this translation? I will confess that as an evolutionary biologist trained in "Western" biology in India and the United States, the answer surprised me perhaps more than most. While this book is written in English (and quite honestly my very postcolonial urban education precludes me from writing a whole book on biology in any other language), I am deeply aware of my inability to translate life and language on the ground in India at every step of the way. I have persevered, but it seems an impossible challenge. Writing in the master's tongue, perhaps I could at least subvert the master's tools and genres? Feminist studies offers generative scholarship in the visionary works of scholars such as Gloria Anzaldúa, Audre Lorde, Cherríe Moraga, Octavia Butler, and Suniti Namjoshi, who have opened up writing as a site of epistemological challenge, political possibilities, and indeed a necessity for survival. Writing opens up new imaginaries, liberatory landscapes, and paths of resistance. As Anzaldúa (1983) powerfully writes in "Speaking in Tongues: A Letter to 3rd World Women Writers":

> Why am I compelled to write? . . . Because the world I create in
> the writing compensates for what the real world does not give me.
> By writing I put order in the world, give it a handle so I can grasp it.
> I write because life does not appease my appetites and hunger. I
> write . . . To become more intimate with myself and you. To discover
> myself, to preserve myself, to make myself, to achieve self-autonomy.
> To dispel the myths that I am a mad prophet or a poor suffering
> soul. To convince myself that I am worthy and that what I have to
> say is not a pile of shit. . . . Finally, I write because I'm scared of
> writing but I'm more scared of not writing.

Inspired by their work, the experiments in this book attempt to destabilize the neat categories of the Western academy, helping us reanimate histories and theories to make the familiar unfamiliar and the unfamiliar familiar. I find it best to describe this with a visual metaphor. Later in this chapter, in developing a naturecultural methodology, I draw on the organic forms of plant thigmotropism (Subramaniam 2014)—the touch sensitivity of organisms, especially plants with tendrils that scale large surfaces through tactile means. Through their thigmotropic tendrils, climbing plants produce a dense braid or weave of organic matter. I have found this an apt metaphor for postcolonial science. Western science when imposed on India was not rejected, but it did not fully "translate" into local contexts either. As the thigmotropic tendrils of Western knowledge scaled Indian cultural and knowledge landscapes, they produced an uneven weave of Western and Eastern knowledge systems. Much was "lost in translation." There are thus many gaps and holes and many dense thickets and knots. A singular challenge has been communicating the many ways in which Western "universal" science re-emerges in India with local inflections, and in deeply uneven and heterogeneous terrain.

To illustrate this, I want to name here the particular concepts, ontologies, and ideas that have been the toughest to translate. I use the terminology that is by now well established in cultural and feminist theories of science, but I want to flag their inadequacy even as I adopt them in the rest of this book. The prefix *bio* (biology, biopower, biopolitics, bionationalism) is perhaps central, at the heart of my disciplinary training in the biological sciences, and the critical focus of this book. In English, the root is the Greek *bios*, which refers to the form or manner in which life is lived, finite and mortal, contrasted with *zoē*, the biological fact of life, infinite and a general phenomenon of life. In outlining the concept of biopolitics and biopower, Foucault argued that in politics in the eighteenth century, Western societies consolidated power into a new medical authority, a scientized discourse of disease, and a new institutional power for science and medicine. Foucault chronicles how the growth of industrial capitalism ushered in new forms of power that constitute life and death, i.e., biopower (Foucault [1976] 1998). This shift in power marked the ascendency of the life sciences, giving them a central role in constituting life, the body, and health. In more recent times, Agamben (1995) has argued that in Western thinking, the distinction between the two roots of "life," *bios* and *zoē*, is increasingly lost, and as a result "life" is more or less exclusively in the realm of the biological dimensions—and there are no guarantees about the quality of life or *bios*. This constitution of a more scientific and medical conception of life does not fully translate into the Indian context. As I explored biopolitics in India, I encountered many dense knots and threads of incommensurability. We can observe this at three different levels. First, in Hindi, many words broadly translate the English *life*, such as *zindagi, jeevan, pran, atma, rahen, josh, jaan*, and *aayushya*, as do the Tamil words *vazhkai, uyire*, and *ayul*. In part, this is because the Indian words continue to incorporate and embody ideas of soul, spirit, energy, and other vitalist notions of life (potentially inhabiting different bodies across generations), ideas long purged from Western definitions. Second, there is a distinct biopolitics of Hindu nationalism, a politics that is incorporated in the body's physiognomy, in biology and in practices of bodily discipline, as well as a biopower that governs populations and governments in the Foucauldian sense. This "biopolitics" is not the same as the Foucauldian transformations in the nineteenth century in the West, but neither is it unrelated; the Indian context embodies a different genealogy of life and time into its conceptions. Finally, we have biopolitics of the

Indian State that engages with world trade laws, patent laws, transnational health initiatives, and rules of the World Health Organization, necessarily adopted for "Western" standards and transnational purposes. The census, biological variables of health and illness, population level indicators, and indices of national health all follow a Western template. Thus, "life" in India inhabits several registers, including untranslatable local understandings, various dimensions of biopolitics of Hindu nationalism, and a Foucauldian unfolding of Western global biopolitics. The politics of life on the ground in India reveals multiple elisions and erasures, and one finds that much is lost in translation.

We see such fissures, mistranslations, and knotty problems throughout the book. I have tried to explicate the Indian contexts whenever possible, but in reality, English does not have an equivalent vocabulary for India, nor the biological sciences the ontologies for these concepts. For example, the multiple sexual identities and ontologies in India, as we shall see in chapter 2, do not translate into the simple politics of "queer," and neither is the term *queer* always intelligible in Indian contexts. Yet I embrace the term because *queer* has emerged as a capacious word to signal not only nonnormative identities but also a method that troubles easy definitions and opens up new constellations of meanings. But such shorthands flatten the diversity and vitality of "life." To adequately translate many of these terms would each take a chapter, and in some cases a whole book. Concepts of time, space, history, linearity, progress, nation, change, origin, death, and body all have different lives in India. In a country where ghosts, souls, and spirit worlds remain alive, and where processes such as reincarnation and transmigration of souls remain deeply revered, Western scientific concepts can never fully take root and thrive. Rather, they live on awkwardly and often violently alongside Indian ways of living in their multiplicities. Many of the fissures and knots of biopolitics in India that haunt this book are contests of such power. In the five illustrative cases that follow this introduction, ideas do not have unidirectional modes; rather, they travel back and forth between West and East endlessly. With each trip, something is lost and gained, translated and mistranslated, resulting in layered genealogies and histories of meaning.

6. BIONARRATIVES: METHODOLOGICAL IMPERATIVES IN WRITING POSTCOLONIAL BIOLOGIES[12]

> Those who do not have the power over the story that dominates
> their lives, the power to retell it, rethink it, deconstruct it, joke
> about it, and change it as times change, truly are powerless, because
> they cannot think new thoughts.
>
> SALMAN RUSHDIE, *One Thousand Days in a Balloon*

If there is one lesson I've learned in the humanities it is that stories are never innocent. Claiming gods and goddesses is a political act. Embracing science as the site of reason or unreason is a political act. We live in political times as we sift through our gods and goddesses of religion and science. Narrating origin stories, adjudicating history, and telling stories of the past, present, and future through the powers of science or divine intervention are deeply political acts. Each enables possibilities often foreclosed through the other.

When I was introduced to the humanities as a biologist, a central insight I had was on the power of narratives and storytelling. But narratives, I have come to discover, are not the sole purview of the humanities. Indeed, science and scientific papers are fundamentally a narrative about nature and the natural world. As Donna Haraway (1989: 5) astutely notes:

> Biology is inherently historical, and its form of discourse is inherently narrative. Biology as a way of knowing the world is kin to Romantic literature, with its discourse about organic form and function. Biology is the fiction appropriate to objects called organisms; biology fashions the facts "discovered" from organic beings. Organisms perform for the biologist, who transforms that performance into a truth attested by disciplined experience; i.e., into a fact, the jointly accomplished deed or feat of the scientists and the organism. Romanticism passes into realism, and realism into naturalism, genius into progress, insight into fact. Both the scientist and the organism are actors in a story-telling practice.

Narratives in this mode aren't about fiction or the fantastical, but any plot—describing an experiment we conduct or a fictional world we imagine—is inherently about narrative. Narratives narrate; they tell a story. How narratives unfold, what is included in and excluded from them, and why, where, and when they are narrated is shaped by convention; in academia, convention belongs in the realm of disciplines. Genres of writing—scientific experiments, social science studies, literary analyses, creative writing—are deeply disciplinary modes of knowledge production. Claims of truth or falsehood, robustness or lack thereof, and rigor or weakness are familiar territory for feminist STS. Feminist and postcolonial STS offer us robust epistemological critiques of science even while passionately insisting on the need for interdisciplinary and contextualized knowledge processes, albeit partial, situated, and contingent (Haraway 1988).

Braided Sciences, Helical Stories, Thigmotropic Knowledge: Toward an Experimental Humanities

Narratives have emerged as a central mode of politics. In chronicling how narrative has moved from literature to politics, John Lancaster (2011) remarks, "We no longer have debates, we have conflicting narratives." As we will see in the five cases in this book, the challenges of scientific knowledge lie not only in the data and their salience but in how scientists and other actors narrate those data in the stories they tell. Three decades of work in feminist STS have highlighted the power of understanding scientific experimentation and knowledge making as narrative. The narratives of scientific papers are not innocent but emerge out of particular commitments to Enlightenment logics. It has led me to imagine an experimental humanities (here an experimental biohumanities) in bringing together empiricism and analytic conventions of experiments from the sciences and the rhetorical and contextual analyses from the humanities.

As a biologist interested in producing knowledge about the natural world, I am compelled by decades of analyses in the feminist and social studies of science that reveal that science is not a value-neutral or unbiased enterprise. Rather, we need to expand the methods and methodologies of scientific training to engage a larger repertoire that a vibrant interdisciplinarity allows. One of the profound lessons I learned in examining the history of biology in *Ghost Stories for Darwin* (Subramaniam 2014) was the richness of biological thought. The sciences are often far more diverse in their theories, opinions, practitioners, and politics, and more contentious and full of debate, than the official histories of science or biology often narrate. Similarly, studies of how science is communicated, how scientific ideas travel and circulate, reveal the inextricable connections among science and larger cultural and social forces. "Science" has also emerged as a site of such epistemic authority that everyone has embraced it—we have Western science, indigenous sciences, Christian science, Vedic sciences, Hindu science, Islamic science, queer science, third-world science, postcolonial science, and so on. In the proliferation of sciences, much is lost and gained; the move to embrace the word *science* is a political project of epistemic authority, and sometimes of obfuscation; the word should always be read with these multiple valences unless otherwise specified. There are many claims to "science," and understanding what science is, who practices it, and why and how science is supported or challenged in various contexts has emerged as a key question of our times.

In exploring these larger questions of science in this book, I engage the analytic powers of biology and the humanities through an examination of narratives. What stories does science tell about itself? How does it narrate biology and nature? How does science get narrated in various contexts? Who tells science's stories, and why and how? I'm interested in the complex and global circulations of scientific narratives. In addition, I'm interested in highlighting the multiple, contradictory, and conflicting narratives of science, by science, and about science. Finally, I'm interested in the stories not told. What are the stories about science that get underplayed, minimized, or erased entirely? To understand the workings of science through an attention to narrative, we need to expand our repertoire of storytelling practices. Versions of the stories in this book have been told elsewhere; yet when the same story is narrated through the theories of feminism, postcolonialism, and STS, something different emerges. That is the power and the purpose behind an experimental humanities.

Two sets of metaphors are striking in our recent understandings of science. First, as our understanding of networks, assemblages, and actor network theory has grown, we talk about science in the plural, and in its relationships with other objects (Luckhurst 2006). Metaphors include fabric

(Wilholt 2013), patchworks (Bauchspies and Bellacasa 2009), entangle-ments (Mamo and Fishman 2013), braided sciences (Mukharji 2016), and the game of cat's cradle (Haraway 1994)—all of which remind us that the tapestries of science and scientific knowledge involve multiple threads, braids and weaving. Taken together, these metaphors remind us that science is not a purified knowledge that is transmitted untouched by social and political forces. Rather, histories, politics, and social structures are deeply woven into the fabric of science and the body politic. It is impos-sible to pull out "pure" knowledge from an intricate weave. Therefore, in discussing science, we need to examine science from all sides as its threads wind and unwind, tangle and untangle across time and space. We need to move from linear storytelling to helical stories as we track the various threads that make scientific knowledge (Gallais and Pollina 1974).

Second, we talk about science using metaphors of movement and mobility. Scholars have discussed contact zones (Pratt 1992), science in motion (Prasad 2008), a moving metropolis (MacLeod 2000), or science as circulation (Raj 2006). These works remind us that science is an evolv-ing enterprise—it moves, traveling across time and space. Postcolonial STS has powerfully documented that science is never exactly the same across contexts (W. Anderson 1998; Roberts 2012; Abraham 2013). Like the thigmotropic tendrils of plants, the fibrils of science seek support structures, and thus come to scale political and social scaffolding in dif-ferent contexts. I use the term *fibril*, from botany, to highlight the organic, living nature of knowledge as it travels and takes different shapes in differ-ent contexts. As science's thigmotropic tendrils build infrastructures in varied contexts, they build on the local terrains and scaffolding and organ-ically create dense knots and thickets of social meaning. Hence, while we interpret these knots as social categories—sex, gender, race, class, caste, sexuality—they are never the same across the world. Tracing science's thigmotropism is important because it has been so central to producing a politics of "difference" (Hamilton, Subramaniam, and Willey 2017).

Taken together, the two metaphors of entanglement and thigmotro-pism align with recent impulses in contemporary STS toward complex stories, toward tracking the entangled fabric of science and society and the migrations and circulations of science and scientific knowledge. Learn-ing to develop such "thick" descriptions necessitates the rich modes of the humanities and the sciences. Indeed, it reminds us of why literature and science have long been important interlocutors even while a disciplinary academy has relegated them to mutually exclusive corners of the modern university.

In using narrative as a mode of the experimental humanities, Anna Tsing's (2015: 38) latest work is provocative in this regard:

> To listen to a rush of stories is a method. Why not make the strong claim and call it a science, an addition to knowledge? Its research object is contaminated diversity; its unit of analysis is indeterminate encounter. . . . A rush of stories cannot be neatly summed up. Its scales do not nest neatly; they draw attention to interrupting geographies and tempos. These interruptions elicit more stories. This is the rush of stories' power as a science. Yet it is just these interruptions that step out of the bounds of most modern science, which demands the possibility for infinite expansion without changing the research framework.

The power of narratives is their ability to make the unfamiliar (scientific studies) familiar (stories) for the nonscientist and the familiar (scientific studies) into unfamiliar (stories) for the scientist. Both force us to locate scientific studies and the enterprise of science within their historical contexts. Scientific narratives in the West draw from Judeo-Christian frameworks (Haraway 1997). The five illustrative cases that animate this book challenge the linear Judeo-Christian frameworks of progress toward greater truth. Rather, they remind us that other narratives exist, ones that travel, evolve, and are endlessly recursive and hyperlinked. Narrative practices in India are a good example. Indian mythological stories are primarily oral traditions. Narrated and renarrated, these tales traveled, mutated and transformed. When written down, they often emerged in multiple versions, the author of each tale imbuing it with a different ethos, and indeed with different story lines and details. Elite versions are routinely challenged by more populist retellings.

Two modes of storytelling practices animate this book. First, as the opening rush of stories suggest, Indian mythology does not follow the linear, Judeo-Christian-inspired storytelling practices of the West. Its many modes are worth highlighting. First, there is no binary good and evil. Indeed, Indian mythology hosts deeply flawed gods, goddesses, and demons. Good and evil can reside in the same individual. As the *Bhagwad Gita* reveals, in the famous exchange between Arjuna and Lord Krishna, the goal of war may not even be justice or righteousness but the fulfillment of duty. Hindu mythology opens up other modes and goals of life and living, not just the good vanquishing evil. The gray areas—those infinite spaces between zero and one, as Kabir's opening epigraph to this book beautifully illustrates—are the spaces of life and living. Second, the boundaries between human and nonhuman, animal and plant, and material and spirit are porous at best. Through transmorphing of bodies, transmigration of souls, and reincarnation, individuals can move across bodies, and their karma and dharma of life travel with them to new bodies and new futures. Third, there are stories within stories within stories. There is no one linear story to be told. Stories may be repeated but always with variations—as a different character is highlighted and new unfolding story utterly transforms the old one. Indian mythological stories are best understood as cyclical or helical stories; we can come back to them again and again, yet utterly anew. A narrative is not a simple logical operation; rather, it is a discourse that unwinds. Each of the five illustrative cases here is a complex case. There are stories within stories that weave together into complex bionarratives of how science and religion are imbricated. Fourth, life is not finite with a clear beginning and end, life and death; it can be reborn in new bodies. Mythological stories follow the same pattern. Recursivity is also a reminder that history repeats itself. Throughout this book we see how old social categories, old histories, and old debates re-emerge in new biological variables, theories, and debates. While rebirth and reincarnation can indeed lend stories their cyclical properties, so can the ghosts caught in the cycles of injustice. The ghosts of misogyny, racism, colonialism, casteism, heterosexism, and other long-enduring historical forces also live on (Subramaniam 2014).

The second storytelling mode that animates this book is more organic—the thigmotropic properties of plants. In *Ghost Stories for Darwin*, I acknowledge my debt to morning glories and the power of their tendrils to latch onto the scaffolding structures around their evolving morphology (Subramaniam 2014). Morning glories will climb a pole in a linear form if that is what is available; they will stay close to the ground and build matted networks through stalks of grass and weeds; or they will rise up as giant and imposing structures as they wind their way around their world. Sciences, like morning glories, are never pure as they travel and interact with other local and contextual knowledge systems to produce a wondrous array of helical stories and thigmotropic knowledges. Projit Mukharji (2016: 27), in his evocative description of "braided sciences," describes knowledge systems as "multi-stranded spools selectively braided together by the presence of the 'entangled objects.' But this braiding in turn also transforms the objects themselves, giving them new meanings and transforming the relations between the two cultural entities being crossed." Like morning glories, science's thigmotropic tendrils are entangled in different biological, social, cultural, and political formations and imbricated in local genealogies, histories, cultures, and politics. Indeed, the braided sciences of South India are different from those of the North. There is no universal story; it is different each time.

Tracing the layered histories through science allows us to see how our narratives of "biology"—what it is and its relation to the world—change over time. At any historical moment, scientific ideas are always multiple, always contested. Indeed, as scholars have reread old scientific texts, other possibilities, sometimes more progressive and imaginative, emerge (Grosz 2011; Hustak and Myers 2012). While particular ideas win out at particular moments, it is important to remember the richness and plurality of scientific thought. Other options, other futures, were always possible.

Finally, the book attempts to embody the multiple narratives and entangled histories of science and religion in its thigmotropic structures as it weaves fiction with nonfiction and multiple but related tales within each chapter. I have incorporated not only inspirations of Indian modes of storytelling but stories themselves. Indian mythological stories have both progressive and regressive elements. While a more virulent form of Hindu nationalism embraces some mythologies in order to purge India of other religions and recast India as a Hindu nation, other possibilities also exist. The opening descriptions of Indian mythologies, and the Avatar stories interspersed between the main chapters, are attempts at presenting an "elsewhere," alternate sites where scientific and religious thought may come together not in the hypernationalist modes but in alternative cadences of Indian storytelling. As you will see, the stories aren't quite religious mythology (if aliens were gods, they could easily be!) but are more in the vein of speculative fiction, deeply inspired by mythology and science.

These experiments within the humanities reveal how historical processes shape a living and breathing science that evolves across time and space. We can now spin old stories anew, unravel and reweave new tapestries of nature, of biology, of the human and the world. The power of the experimental humanities is in its interdisciplinary methods and methodologies—to experiment, explore naturecultural worlds, while at the same time being deeply cognizant of what the humanities offer us: an attention to the histories of terms, theories, language, frameworks, and contexts. An experimental biohumanities refuses the idea that the cultural world does not matter to the sciences or the natural world to the humanities. There is critical value in studying them together.

7. AVATARS FOR STOLEN DREAMS

> One might say that the epistemic story of imperialism is the story of a series of interruptions, a repeated tearing of time that cannot be sutured.
>
> GAYATRI SPIVAK, *A Critique of Postcolonial Reason*

I return to Modi's claims of Hindu mythology as evidence for Vedic plastic surgery and genetic sciences with which I began this chapter. While it may be amusing, it opens up the possibilities of new imaginative worlds, albeit not in the supremacist mode that Modi often invokes. Whether one reads these stories as fact, fiction, myth, metaphor, or allegory, they can reveal a stunning and pluralistic imagination of the natural and cultural, of gender, sexuality, and kinship. In these worlds, the newer conceptions in Western science that create the binary realms of nature and culture, human and nonhuman, and scientific and spiritual mingle to create nonbinary imaginations of naturecultural, human/nonhuman, and spiritual scientific worlds. If we think of these stories alongside what we may call biology, the body, the flesh or matter, these stories are generative and allow new naturecultural possibilities. Trees communicate with each other and other inhabitants on Earth. Human and nonhuman communication is rampant. Reproduction is not always grounded in sex, heterosexuality, or heteronormative reproduction. Sex and gender are mutable. New bodies can emerge from tissues, through desire, through necessity, through longing, through invocation to the gods. Reproduction is possible in bodies of either gender or indeterminate genders, through nonmaterial ghostly worlds and magical spectral creatures. Reproduction is possible across vast spatial scales. Pregnancy can be entirely unmediated and possible outside of the ontologies of bodies. The metaphoric possibilities of epigenetics are echoed in stories of fetuses in porous wombs, imbibing their cultural surroundings.

An experimental humanities helps us understand that there is no linear link between stories, their narrative power, and their interpretation. Stories are always interpreted through politics. None of the stories automatically unfold into the politically progressive or regressive. Indeed, stories of queer reproduction can reinforce heteronormativity, epigenetics can reinforce patriarchal ideals, and human/animal interactions can reinforce essentialist ideas of species. What excites me about these stories is not that they are an a priori site of radical politics but that they eschew a singular hegemonic narrative and open up the possibilities of other worlds and other ways of living as legitimate and possible—ones that make you sit up and ask "why not?" and ones that make you wonder about the varying contexts that produced such breathtakingly diverse renderings.

What is particularly striking (and what I believe propels Hindu nationalists) is that so much of the mythological imagination that Modi and other Hindu nationalists narrate is now in fact the focus of the cutting edge of contemporary biosciences. Trees do communicate with each other, as do animals; the womb and fetus do not exist in isolation; reproductive technologies have expanded the possibilities of kinship; and indeed bodies are profoundly mutable. Hindu nationalists use these modern studies to bolster the veracity of mythological stories. Claiming a robust mythoscientific corpus, they imagine a thoroughly modern, material, global, and scientific Hinduism. They do not rethink or reimagine science. Rather, they displace a "modern" science from the West and reclaim it within the borders and histories of ancient India, even while they claim the "cultural" modes of a Hinduism steeped in casteist, heterosexist, and patriarchal norms. Unlike other religious fundamentalisms, modern Hinduism has produced not a scriptural fundamentalism but a political nationalism through a melding of science and religion.

Finally, one of the striking features of India is its syncretic pragmatism. While Hindu nationalism has grown in power and strength, religious followers still visit shrines of saints and gods across religious lines. Some sites are famous for fertility, others for the health of the heart, others for wealth, others for warding off evil, and one even for successfully securing a visa. While ostensibly practicing one religion, pragmatic individuals may acknowledge the powers of various religious sites. Similarly, we see deep pragmatism around health practices. With multiple practitioners of medicine, families routinely will visit doctors of allopathy, homeopathy, Ayurveda, and Unani medical systems. These approaches are not seen as contradictory. In a land of multiple religions, time warps, and polytheism (Hinduism has 33 million gods, according to one count) (Dasa 2012), there is no singular epistemology or ideology that limits that pragmatism. Mythological tales flow into scientific tales, the natural and supernatural mingle, and science and religion cohere into pragmatic living.

Science and religion are not oppositional. They are something else— tools, allies, synergies, partners, symbionts, challengers, colluders, or syncretic collaborators. The contours of modernity in India expose the polyvocal imbrications of science and religion.

As the five illustrative cases unfolded, what emerged repeatedly were the politics of crossings—of exchange, hybridity, and translation. The vibrant field of trans studies, with its treatments of crossing, shifting, traveling, pivoting, thwarting, has undoubtedly inspired my thinking. Because of the books' focus on India, I was drawn to the figure of the avatar in Indian mythology, and its trans politics, as a potent epistemological tool to understand postcolonial formations of science and religion.

Avatars, a central concept within Hindu mythology, suggest the divine coming down into human and animal forms. The root of *trans-* shares a commonality with a Sanskrit root, *tara*, both meaning "cross." *Avatara* has to do with the crossing of deities—their descent to Earth (Osuri 2011). In the famous mythological tales of Dasavatar, the stories depict the god Vishnu's ten avatars. In these stories, problems caused by powerful demigods leave the Earth in chaos, and Vishnu, embodying various human/nonhuman forms, descends to Earth to bring peace again. In computer gaming today, avatars have come to represent online identities of individuals. At the heart of avatars is a disassembling of identity and form, and its subsequent assembling into another. Avatars are an expression of a general condition of entanglement and highlight the entanglement of the divine with earthly life. Even the word *tantra*, which I've always associated with eroticism in the West, has a different valence in India (Dhamija 2000). The Sanskrit term translates to "weave, loom, and warp," suggesting the interweaving of traditions and teachings as threads into a text, technique, or practice. The Hindu god Vishnu is called *tantuvardan* or "weaver" because he is said to have woven the rays of the sun into a garment for himself (M. Chatterjee 2000). Similarly, the concept of karma is a force that works across/trans individuals; every act is a karma of various people.

I use *avatar* rather than *trans/national* because with its Indian resonance, it offers more—of translation, traversing, transmigration, transmutation, transnational, transmogrification. Within Indian storytelling practices, one doesn't "cross" from one ontological entity to another. Rather, souls, spirits, and ghosts can travel, cross, and return. They can also be in multiple places at once. Like the avatars of Hindu mythology, the various avatars of bionationalism in this book take different forms because they are intervening in different problems to offer different solutions. However, the scientific world, like the *devalok*, where the gods reside, lives on. Local avatars of bionationalism do not appear to challenge or transform the scientific or godly worlds. Mythology and avatars are also useful because they resist the idea that everything is always translatable. As the case studies reveal, avatars can be unsuccessful, be partially successful, or spawn new problems. These spaces of ambivalence, contradictions, and possibilities are important spaces that reveal the need for theories of "trans"/crossings.

Avatars and their trans politics also signal how modern bionationalism traverses many borders into geographic, temporal, and disciplinary crossings. The transnational emerges as a critically important site for Hindu nationalism as an active and passionate diaspora spreads the reach of Hindu nationalism abroad. Indeed, "instead of encouraging a sense of world citizenship, the transnational experience seems to reinforce nationalist as well as religious identity" (van der Veer 1994: xii). A "long distance nationalism" (B. Anderson 1992) mobilizes and unifies a diasporic community on a "world mission" to propagate the Hindu notion of the world into a united and single global family (Andersen and Damle 2018).

Taken together, these five cases represent very particular and deliberate forms of bionationalism—where science and religion come together in specific formations in order to propel a particular vision of the nation. The stories do not represent the triumph of Hindu nationalism; indeed, secular nationalism is an important part of many of these stories. Rather, what they expose is how biopolitics has become a critical frame through which contemporary state and nonstate actors act. These are the new avatars of bionationalism, each avatar representing a site of heterogeneously imbricated formations of science and religion. These avatars are techno-religious objects. They are particular and peculiar formations—institutional and instrumental—in the quest for a modern global India. They thrive because of the confluence of particular historical, economic, cultural, political, and global contexts that cohere to shore up India as a "wounded civilization" and Hindu nationalism as its biopolitical power and savior. The goal is for India to have its rightful place in the global pantheon of superpowers.

As we see a growth in scholarship of feminist, postcolonial, transnational, and indigenous STS, we have seen the rise of many salvation narratives for science and STS. Scholars have lamented that Western enlightenment has arrived at the winter of its discontent; others exhort the third world to eschew Western models and embrace its indigeneity. Yet a perusal of postcolonial STS and Indian nationalism suggests that knowledge claims do not automatically render progressive worlds. To enable progressive and liberatory worlds, we need progressive and liberatory politics that are attuned to the oppression and violence of our histories. India and other third-world nations may claim their own enlightenment, their own modernities. Similarly, the West may never have been modern (Latour 1993) and may have always lived in different registers of rationality (Chakrabarty 2000). Giving up these easy tropes of the West and the East, of modernity and nonmodernity, reveals the need for new narratives for planetary salvation (Meighoo 2016). To understand India is to also appreciate its many genealogies and its lush imaginations of the real and unreal, the rational and irrational, the natural and preternatural, of this world and others, and learning to see, feel, hear, touch, listen, and mingle with fellow creatures in this world and others. To understand how India got from there to here requires us to embrace the phantasmagoric lives of science and religion. To embrace the many stolen dreams, fantastical narratives, storytelling traditions, and progressive possibilities that were appropriated over the long histories of colonialism and conquest, and over the violence committed in the name of caste, gender, sexuality, nation, science, and religion. We open ourselves to the possibility that in embracing such netherworlds—their dizzying spirits, their disavowal of the strictures of rationality, reason, and civilizational logic—we may dream and build new avatars to fulfill our forgotten dreams for more vital, vibrant, and just futures.

AVATAR #1

The Story of Uruvam

Uruvam extended its abundant tentacles across Kari. Its multisensory skin turned translucent as it sampled the environmental conditions. Waves of color roiled through its agile body as it absorbed the information. Once done, Uruvam retracted its tentacles into its body and transformed itself into a black octahedron. Uruvam was the node, or Kanu, of *form* for the experiment. It was capable of morphing itself into infinite shapes. The avatars joked that Uruvam, constantly transforming into intricate patterns and shapes, had a geometric soul. As it processed its newly absorbed information, Uruvam contemplated the landscapes on Kari, noting that the skies were a little less blue. There were more swirling clouds hovering over the planet than in the past. Uruvam morphed into a purple cuboid and reflected on the disappointing news. How much it had tried to get earthlings to appreciate the joys of forms and morphologies that earthly biology made possible! How wondrous was a biology grounded in the constantly mutating and recombining biology of DNA on a planet whose environment was forever changing. The current iteration on Kari fascinated the Kankavars because of the surprising evolution of humans, who called themselves *Homo sapiens*. Things were going well until certain groups of humans crowned themselves judge and jury of all life on Kari. They had developed new concepts like "beauty," but rather than learning to appreciate the multiple forms of beauty everywhere, beauty became a singular and universal concept, and *Homo sapiens* organized and rewarded those earthlings who followed a very narrow conception of form and beauty.

Uruvam was not happy. What irked it more than anything else, it said, was earthly mathematics. Why couldn't they count past two? They'd discovered zero and infinity and everything in between, yet they could not give up their love for thinking in dualisms. Uruvam had created and introduced so many bodily morphologies and inventive body parts—some critical, others suspiciously useless—each of which displayed endless variation, yet humans could not see it. With time, earthlings had discovered some variations. Some plants and animals embraced spectacularly intricate and outrageous morphologies. They preened, displayed, reveled in, and paraded their extravagant structures. The aberrant species was the human, which still had failed to grasp the rhythms of carbon-based evolution.

Uruvam had created groups with a multiplicity of religions and religious icons—virgin birth, multiple heads, multiple hands, human-animal hybrid gods, human-plant gods. But *Homo sapiens* inevitably chronicled that immense diversity into two groups—gods and goddesses, gods and demons, Devas and Asuras, humans and nonhumans, animals and plants. Only two. Two sexes, male and female; two genders, masculine and feminine; two sexualities, homosexual and heterosexual. They had invented other terms—race, caste, ability, culture, illness—and again, despite many options, *Homo sapiens* always reduced it to black and white, upper and lower caste, able bodied and disabled, high and low culture, healthy and ill. It was always an "us" and "them." And if you were unfortunate enough to be a "them," you were relegated to a separate category, as the "other," deemed inferior and not worthy of any respect or consideration.

Uruvam introduced many technologies—body enhancement, genetic engineering, body sculpting, sex change, sexual transformation, body morphing, chimeric development, gender enhancement, pansexuality machines, hybridization, skin grafting, organ development—but it seemed that whatever it introduced, the humans only used it to mold themselves into an increasingly narrow conception of beauty. How these oppressive humans had tortured and vilified those "other" creatures on Kari—like they were put on Kari to be served! Some *Homo sapiens* had noticed this and been outraged by it; they pointed it out, had written about it, and indeed devoted whole academic fields to it. But the mathematics of "two" would not budge! *Are all my wonderful imaginations are going to be ultimately reduced to this,* Uruvam sighed.

Uruvam had planted the seeds of a plenitude of modes of life—decomposition, recycling, reincarnation, transmigration of souls, spirits, ghouls, zombies, and ghosts. Over time, Uruvam fretted that any group that believed in spirits or ghosts was quickly written off as "primitive" and sidelined as unscientific and unwise. Uruvam had anticipated even this and sowed the seeds for other theories. Some *Homo sapiens* had uncovered clues to understand the natural cycles of carbon biologies—the decaying and decomposition, the possibilities of recycling and composting. But others, alas the majority, could still not see that all new beings were carbon matter and must therefore come from the disintegration of old carbon beings. Why could humans not appreciate that the planet was organized around cyclicity, not linearity?

Uruvam remembered the many experiments. The duckbill platypus, when it was discovered, looked to the British like an animal with the bill of a duck attached to the skin of a mole. Uruvam was amused when the British wondered if Chinese sailors were playing a joke and had stitched this hybrid creature together. But these moments of skepticism never lasted. With time, all the anomalies and exceptions were neatly explained away. With the need for singular theories, humans used criteria like parsimony and elegance to weed out inconvenient complexity. Ultimately, all of Uruvam's wondrous, playful, and creative experiments were neutralized by biological and scientific theories. Uruvam had descended as several famous scientific avatars. These avatars espoused a humanism to counteract the increasing dehumanization of scientific thought. They espoused the majesty and beauty of the planet and its life forms, the infinite universe, and the enchantment of the wondrous world around. While the avatars were initially hailed as visionaries, the enchantment, humility, and sense of wonder was eventually forgotten, and the disillusioned souls floated back to Avatara Lokam. They and their visions were once again replaced by a singular certainty and objectivity. Diversity of thought, like diversity of form, Uruvam lamented, was always contained by the dictates of theory.

The final stages of Uruvam's plans were unfolding. Uruvam started working with Amudha, and together wondrous morphologies were unfolding on Kari. Would wonder and beauty return to Kari?

CHAPTER ONE

Home and the World

The Modern Lives of the Vedic Sciences

After Independence, India's hydroelectric dams—erected as they
were to a secular faith—came to be known as Nehru's "new temples." Such projects were to ensure India's future. Today, many recognize that the wings of modernity were not powerful enough to
bear the huge nation aloft. . . . Traditionalists have countered the
modernizers with their own Enlightenment.

RAJNI KOTHARI, "The Indian Enterprise Today"

India is not, as people keep calling it, an underdeveloped country,
but rather, in the context of its history and cultural heritage, a highly
developed one in an advanced state of decay.

SHASHI THAROOR, "Globalization and the Human Imagination"

NARENDRA MODI WAS SWORN IN AS THE FIFTEENTH PRIME MIN-
ister of India on May 26, 2014 (Burke 2014). For the first time in Indian
history, Modi's party, the Hindu nationalist Bharatiya Janata Party (BJP),
won a landslide election with enough votes (292 seats out of 543) to form
a government, while the opposition, the Congress Party, was reduced to a
historic low, winning only 44 seats. The victory of the BJP is seen widely
as the victory and success of Narendra Modi, former chief minister of the
state of Gujarat (*Economist* 2014). Modi ran a powerful media campaign
that was organized around the message of "development,"[1] tapping into
voters' frustration with the previous government and anger against its

corruption (*EPW* 2014; Jayasekara 2014). "Development nationalism," a slogan Modi developed, emerges from his purported successes as the chief minister of Gujarat and his so-called Gujarat model of development. While soundly criticized and refuted in some circles, the Gujarat model is also touted as the model for India, a claim that brought Modi and his party to power in 2014 (Hensman 2014). His surging rhetoric during his campaign promised to lead the nation to new heights of prosperity through large-scale development efforts. Development was central to the BJP's platform, and Modi relied heavily on notions of restoring Indian/Hindu tradition to win votes while still promoting an agenda to "modernize" India.

One needs to understand Modi's victory sixty-seven years after India's independence, both in what people saw as a failure of previous (secular) governments and in the steady rise in religious nationalism. Many share Rajni Kothari's (1989) sentiments that India has failed to realize and achieve its potential. With the founding of independent India began a project of imagining India, an India that fulfilled the dreams and hopes of an independent nation and a free people. Yet over the last three decades, the rise of religious nationalism has been steady and unmistakable, and the very definitions of secularism and democracy have shifted. Rather than disavow either, religious nationalists have redefined both—secularism as tolerance and democracy as majoritarianism (Vanaik 1997). Thus, they argue that while the presence of religious minorities should be "tolerated," the majority Hindus should define and govern India. Religious nationalists imagine a Hindu India for a Hindu people. While Modi toned down his militant brand of Hindu nationalism during the election campaign, his history of militant activism, and particularly his role in the 2002 Gujarat pogroms, worries many.[2] What is at stake is who gets to define and imagine India.

The growth of religious nationalism in India over the last few decades has been unmistakable. In 1998 the BJP came to power in India. After thirteen months, the coalition government was toppled, but the next election brought the BJP (in a coalition government again) back to power. The political success of the BJP draws on two other Hindu nationalist movements—the Vishva Hindu Parishad (VHP), an organization of religious leaders, and the Rashtriya Swayamsevak Sangh (RSS), a militant youth organization (van der Veer 1994). The Hindu nationalist program stresses *Hindutva*, or Hinduness. Hindu nationalists have successfully tapped into the overall discontent of Hindu Indians (economic, social, and cultural) and transformed it into a problem about religion and the brand of secularism and democracy India's founders had envisioned (Vanaik 1997). In 2014, for the first time in its history, the BJP won a majority in Parliament. It is noteworthy that the party does not have a single elected Muslim representative in a country that houses the world's second largest Muslim population (Vishnoi and Chishti 2014).

In this chapter, I explore how women, culture, and development have shaped and been shaped by the rise of religious nationalism in contemporary India. Here is a reinvention of India's past in the orientalist traditions of invoking a grand and ancient Hindu past. The traditionalists, as Rajni Kothari (1989) suggests, have countered the modernist project with their own enlightenment, one that reinvokes the grand Vedic tradition. The tradition invoked is one that is scientifically and technologically advanced—a tradition that anticipated the development of, and is thus in harmony with, modern-day science. This new imaginary homeland is used to develop a blueprint for the home and the world, public and private culture, the nation and the individual. The coconstruction of the public and private spheres through a common ideology of *Hindutva* is a particular form of the reinvention of India. The secular, inclusive ideals many grew up with have been literally recast. What is particularly significant is that India as a nation did not exist before 1947. The Indian subcontinent has been home to a multitude of religions for centuries, many emerging from within its own soil. Yet the claim for an "authentic" past is, for Hindu nationalists, an exclusively Hindu past. Such reconstructions mean that what was once made can now be remade; one history can be replaced by a revisionist history.

The project of nation building taps into the syncretic religious ethos in India. A case in point is the launch of India's first Mars orbiter, *Mangalyaan*, in 2013. On the day before the launch, Chairman K. Radhakrishnan of the Indian Space Research Organization (ISRO) offered *pujas* (prayer ritual) to Lord Venkateswara (a form of Vishnu) at the popular Tirupati temple. His predecessor had done the same before similar launches (Patrao 2013). Religion and science are not seen as contradictory but are both deeply woven into the fabric of India (Thomas 2017).

In this chapter, I further develop the ideological basis of Hindu nationalism. Next, I explore the symbolic and ideological role women play in the nationalist reinterpretation of Indian culture. Finally, I examine religious nationalism's use of science, technology, and development. What characterizes this archaic modernity, I argue, is a confluence of masculinity, science, technology, development, and militarization. Despite the rhetoric of angry, vengeful goddesses, women are relegated to the private sphere as vehicles for nurturing generations of Hindu men, who through their masculine agency will work for the recovery of a Hindu India (Pant 1997). To underscore how archaic modernities work in the nationalist imagination, I examine Vaastushastra, the ancient Indian material science, as a case study of a modern-day Vedic science that seamlessly merges the archaic and the modern. In exploring the repackaging of this ancient tradition as a modern material science, I illustrate how archaic modernities function. The confluence of the archaic and the modern works as a metaphor for the home and the world and also as a larger metaphor for nations and nationhood.

THE ARCHAIC AND THE MODERN

> For who among us, after all—white or nonwhite, Western or not—
> is not always caught precisely in the space between "inherited tradi-
> tions" and "modernization projects"? And where else, how else, do
> "cultural interpretations" come from—"theirs" or "ours", local or
> global, resistant or complicit as the case may be—other than from
> the spaces between the two.
>
> FRED PFEIL, "No basta teorizar"

What, then, is the nature of the modern encounter between science and religion in India, and how are the lives of women and the constructions of gender shaped by these intersections? At its independence and creation in 1947, India embarked on a modernization project—a Nehruvian[3] industrial model of development. Western science emerged as "the reason of the state." As Nandy (1988: 7–8) explains:

> This expectation partly explains why science is advertised and sold
> in India the way consumer products are sold in any market economy,
> and why it is sought to be sold by the Indian élites as a cure-all for

the ills of Indian society. Such a public consciousness moves from one euphoria to another. In the 1950's and 1960's it was Atoms for Peace, supposedly the final solution of all energy problems of India; in the 60's and 70's it was the Green Revolution, reportedly the patented cure for food shortage in the country; in the 70's and 80's it is Operation Flood, the talisman for malnutrition through the easy availability of milk for every poor household in the country.

Development—like science—has been central to the modernizing mission of India. The development of India through large-scale industrial growth, hydroelectric dams, agricultural development, and militarization has been central to the policies of previous secular governments as well as the current nationalist imagination. India has been pluralistic about religion, but science and development have been protected spaces in politics. Critics and protesters of development have enumerated the profound, often irreversible, consequences of development through the disenfranchisement and dislocation of peoples and the costs to the environment. Religious nationalists condemn the critics, questioning their patriotism and love of country (Rawat 2000). Similarly, while postcolonial critics have questioned the effectiveness and appropriateness of the particular forms of science and development institutionalized in India, these critiques have largely been ignored by past secular governments and, more recently, by religious nationalists. At first glance, religious nationalists seem to have a critique of the harsh impact of colonialism. However, on closer examination, it appears that religious nationalists embrace some of the more regressive elements of both science and religion, a selective record of history, and a promise of a future that serves only the Hindu elite.

WOMEN, CULTURE, NATION

In response to Christian missionaries and orientalist scholarship, the consolidation of Hinduism and the move to a modern/scientific Hinduism (Oza 2007) began especially with thinkers such as Swami Vivekananda (1863–1902) and Swami Dayananda (1824–1883), central figures in Hinduism. The former set out to "modernize" Hinduism by organizing a disparate set of traditions through a systematic and scientific interpretation of the Vedanta (the Upanishads and the tradition of their interpretation) (van der Veer 1994). India's first modern swami and missionary to the West (Nandy 1995), Swami Vivekananda created a nationalist discourse that is central to Hindu nationalism in all its versions, including the RSS/BJP/VHP brand of Hindu nationalism (P. Chatterjee 1989).

Through nationalism and the creation of the "modern" Hindu, there began a campaign to reshape and recast the role of women (Sangari and Vaid 1989; Amrita Basu 1998b). Hindu nationalism turned women into a national symbol to create the motif of divine feminine power, or Shakti. The myth of Shakti was invoked again and again. The nation, now female and a "motherland," came to symbolize both the powerlessness of the colonial subject as well as "the awakening conscience of her humiliated (Hindu) sons" (Bagchi 1994: 3). The invocation of Devi—the goddess with her garland of skulls, standing on a supine male Shiva—represented both the protectress and the sacred domain to be protected from alien violation (Bagchi 1990). Yet against this image of female power, Hindu leaders relegated women to the domestic sphere and assigned men the role of wage earners. For example, Vivekananda believed that women should not be educated in the modern sciences but should achieve fulfillment within the family (Jayawardena 1988); Dayananda endorsed women's education, but only for a more disciplined child-rearing process (T. Sarkar 1994); and Gandhi espoused the "complementarity" of women and men, with women in the home and men as wage earners (Joshi 1988). Despite other progressive genealogies of gender roles in India (explored later in the book), it is these essentialized binary differences between men and women that emerged in the public and private spheres. The recasting of women as "authentic" was seen as one of the surest signs of the superiority of the East, trapping women in the nation-building process in particular ways (Bagchi 1994: 5). Women are at once mythologized and empowered yet subjugated and disciplined. In this archaic modernity, there is a renewed Hindu masculinity, a rhetoric of symbolic female power that in reality perpetuates the redomestication of women.

In this emerging militant Hindu nationalism, women were limited to three possible roles: heroic mother, chaste wife, or celibate warrior (Sethi 2002; S. Banerjee 2006). Hindu nationalists in their rise have benefited from strident women leaders and have used women leaders in their larger objective of isolating and vilifying the Muslim community. As Uma Chakravarti (1993) demonstrates in her foundational work on gender and caste, upper-caste women became complicit and invested in the caste system because they saw its benefits, even while Brahminic patriarchy controlled their sexuality to ensure the "purity" of upper castes. The woman becomes a key site, her honor tied to that of the honor of the Hindu nation and her family. She is the *matrishakti*, victim and victor at the same time (Singh et al. 2014). In her analysis of three of these figures—Vijayraje Scindia, Uma Bharati, and Sadhvi Rithambara—Amrita Basu (1996) traces their militancy and the striking fact that they are all single (one is widowed; the other two are single, celibate monks, or *sanyasins*). Their chastity underscores their iconic status and their deep association with Hindu spirituality, liberating them from Hindu women's usual ties and obligations to their families (Amrita Basu 1996). As Mridula Singha, national president of the BJP women's organization, says, "for Indian women, liberation means liberation from atrocities. It doesn't mean that women should be relieved of their duties as wives and mothers" (quoted in Amrita Basu 1996: 71). The celebration of this representation of Hindu woman as "ideal" happens alongside the representation of Muslim women as "inherently atavistic" (T. Sarkar 2008; Shandilya 2016). The Hindu nationalist movement elides women's personal self-defense with national self-defense, transforming women into symbolic border guards that help deepen and regulate the boundaries of Hindu and Muslim communities (Sehgal 2015).

As others have argued, this resurgence of Hindu nationalism is a resurgence of Hindu masculinity, a reaction to the "effeminization" of Western colonialism and orientalism (see, for example, Jeffery and Basu 1994). Science and religion have proved to be two powerful tools through which religious nationalists have imagined and engaged with this project of masculinization. Religious nationalists' vision of religion and Hinduism parallels their science policy and is decidedly militaristic and violent. Indeed, Hindu nationalists departed from the pluralistic vision of India's founders.[4] They feared that Mahatma Gandhi's "effeminacy" would bring about the further "emasculation" of Hindu men, and this fear culminated in a Hindu nationalist assassinating Gandhi in 1948. Since the rise of Hindu nationalism, there has been an increase in violence against religious minorities and the destruction and demolition of churches and mosques. Violence, dominance, and an increasingly militaristic policy have marked religious nationalists' quest for a Hindu nation.

Religious nationalism breeds sectarian and communal politics, and inevitably women become markers of the community and synonymous with culture (Butalia 1999). The important work of Urvashi Butalia, Kamla Bhasin, and Nivedita Menon on women and the partition of India and Pakistan point to how the twin discourses of patriarchy and religion use women's sexuality to define notions of honor and shame for the family, community, and nation. The embrace of women's sexual honor and shame have long and violent histories in the name of both religion and culture (Menon and Bhasin 1998; Butalia 2000). Like other nationalisms, Hindu nationalism implicates women in its growth. Contemporary religious nationalism reinvokes Shakti and the power of women, albeit in strategic ways. Primarily antiminority, especially anti-Muslim, in their stance, women leaders herald the great power of women in traditional Hinduism. The president of the all-India BJP women's organization, Mahila Morcha (Women's Front), claims that "In the Vedic era, the status of women used to be much higher than it is today. . . . After the Muslim invasion all of this changed: Hindus were forced to marry off their daughters at much younger ages, they adopted seclusion, and women's role in public life declined" (quoted in Amrita Basu 1998a: 172). During the rise of the BJP, women such as Sadhvi Rithambara and Uma Bharati were formidable figures "projecting themselves as victims of Muslim men." By displacing male violence onto Islam and Muslim men, they claim that their "motherland was being raped by lascivious Muslim men," and they goaded Hindu men to regain their masculinity by violence against Muslims (Basu and Basu 1999). For example, during the Ram Janmabhoomi movement in Ayodhya, which sought to demolish an ancient mosque, the Babri Masjid, and replace it with a Hindu temple because the site was claimed as the birthplace of the Hindu god Rama, Uma Bharati is quoted as saying:

The one who can console our crying motherland, and kill the traitors with bullets, we want light and direction from such a martyr, we want a Patel or a Subhash[5] for our nation. . . .

When ten Bajrangbalis[6] will sit on the chest of every Ali, then only will one know whether this is the birthplace of Ram or the Babri Masjid, then only will one know that this country belongs to Lord Ram. (quoted in Nandy, Trivedy, and Yagnik 1995: 53)

Once in power, the BJP sidelined these women, assumed a more moderate stance, and "channeled [women's] militancy" by sending "women back to their homes" (Amrita Basu 1995).[7] In October 1998, the BJP unsuccessfully introduced a Hindutva plank for the national education conference, which included compulsory housekeeping classes for girls. Again, this "domestic science" invokes the traditions of domesticity with the modernism of science. The rewriting of Indian history and the "saffronizing"[8] of Indian education are actively in process (Panikkar 2001; Raza 2014; Goswami 2017). Consider the Vidya Bharati paper at the State Ministers' Conference on Education in September 1998. It glorified motherhood and named the woman's primary responsibility as the home and the "turning out of good Hindu citizens." Stressing images of Sita and Savitri,[9] the paper advocated that women be obedient and selfless in order to take care of their husbands and family. The practice of *sati*, child marriage, and notions of caste and purity of blood are justified and shown as proud elements of Indian culture (Taneja 2000). The redomestication of women through the power of religion and science is at the heart of this archaic modernity. Amrita Basu's analysis of the Hindu right's rhetoric on the role of Hindu women suggests that the central message is the importance of devotion to their families—and the dangers that await those who refuse to conform. Juxtaposing Indian values with Western ones, they emphasize the links between women's reproductive and social roles (Amrita Basu 1998a, 1998b). Such sentiments fuel social rage and movements such as "love jihad" and "Romeo Jihad," which allege orchestrated campaigns in which Muslim men feign love for non-Muslim women, especially Hindu women, in order to convert them to Islam (Punwani 2014).

In this archaic modernity, the past is glorified and selectively reinvoked to achieve current political purposes. While the top ranks of leadership in the Hindu nationalist movement are filled by men, women played a significant role in its rise. Furthermore, religious nationalists embrace consumerism, globalization, and capitalism. As Tanika Sarkar argues, contemporary religious nationalism does not deny the privileges of consumerist individualism to its women. She persuasively argues that it "simultaneously constructs a revitalized moral vision of domestic and sexual norms that promises to restore the comforts of old sociabilities and familial solidarities without tampering either with women's public role or with consumerist individualism. . . . Patriarchal discipline is reinforced by anticipating and accommodating consumerist aspirations" (T. Sarkar 1994: 104).

Science, Masculinism, and the Bomb

> It had to be done, we had to prove that we are not eunuchs.
>
> BALASAHEB THACKERAY, outspoken Hindu sectarian leader, after the
> nuclear tests in Pokhran (quoted in Swapan Dasgupta 1998: 8)

Leaving aside the advocacy of "domestic science," a more violent nuclear science has been invoked by the ruling BJP to bring about a resurgence of Hindu masculinity, further marginalizing women as sustainers and reproducers of Hindu families. Nowhere was this more apparent than in the euphoria that followed the testing of the nuclear fission bomb in Pokhran in 1998, allowing nationalists to celebrate India's revived masculinity (Chengappa 2015). Swapan Dasgupta (1998, emphasis added) of *India Today* writes:

> Vajpayee has released a flood of pent-up energy, generated a mood of heady triumphalism. He has kick-started India's revival of faith in itself. To the west, the five explosions are evidence of *Hindu nationalism on a viagra high*. The tests lifted up a mound of earth the size of a football field by several meters, and one of the scientists is recorded as saying, "I can now believe stories of Lord Krishna lifting a hill." . . . To Indians, it is evidence that there is nothing to fear but fear itself. Pokhran is only tangentially about security. *Its significance is emotional.* The target isn't China and Pakistan. *It is the soul of India.* . . . The mood is euphoric.

This so-called triumph came at the cost of a military budget of USD $10 billion (an increase of 14 percent over the previous budget), twice the amount spent on education, health, and social services combined, in a country with a female literacy rate of 36 percent, where women earn 26 percent of men's earnings, and where there are 927 women for every 1,000 men in the population (Basu and Basu 1999). After Pokhran, it was estimated that the cost of the nuclear-weaponization program was equivalent to the cost of primary education for all Indian children of school-going age (Raman 2000). Such is the price of military nationalism. As Arundhati Roy (2000: xxiv) writes in *The End of Imagination*, "'These are not just nuclear tests, they are nationalism tests,' we were repeatedly told. This has been hammered home, over and over again. The bomb is India, India is the bomb. Not just India, Hindu India. Therefore, be warned, any criticism of it is not just anti-national, but anti-Hindu." While continuing most of the science policies begun under secular governments, what marks the BJP is its nuclear policy. There have been systematic shifts in budgetary allocations to favor military and nuclear research at the expense of agriculture, health, medicine, and a general science education (Taneja 2000). It is not accidental that despite India's nuclear capabilities for the previous twenty-five years, it was the Hindu nationalists who defied the world to test the ultimate destructive weapon of Western science, the fission bomb, in Pokhran in 1998. Indian nuclear scientists and policy makers form an all-male club. These scientists, co-opted by the nationalist spirit, are India's new heroes. As Balasaheb Thackeray eloquently summarizes, for the Hindu nationalists, the bomb proved India's masculinity. Since forming a majority government in 2014, the BJP has worked further to strengthen India's military capabilities. On June 4, 2018, India successfully tested its long-range ballistic missile Agni-5, the sixth successful test of the missile (*Hindu* 2018).

In this archaic modernity, women and woman-power come to symbolize what is allegedly a triumph of masculinity. The bomb, this ultimate destructive weapon, was christened Shakti after the goddess of power and strength. After the testing of the bomb, the VHP began plans to construct a temple at Pokhran, dedicated to the goddess Shakti, to commemorate the tests; it was named Shakti Peeth (altar of Shakti). The VHP general secretary invoked women again in suggesting that this was an ideal location for Shakti Peeth, as "Baba Ramdev[10] is worshipped here for the reforms he brought about in the society, especially for waging a movement for the protection of women" (*Indian Express* 1998). Some nationalists also proposed that the sacred soil of Pokhran be carried in sanctified vessels across the country in a set of jubilant *yatras*[11] (T. Singh 1998).

The first anniversary of the Pokhran tests was declared Technology Day, and the human resources and development minister, Murli Manohar Joshi, while laying the foundation for a new technology forecasting center, stated that "Pokhran and all our scientific endeavors have brought glory to India" (*CNN* 1999). The nuclear tests in Pokhran have come to symbolize the success of religious nationalism by proving Indian power and might, its strength and scientific capabilities. Central to this model of industrial science and development is a resurgence of Hindu masculinity and the increasing marginalization of women into the cultural role of the sustainers and reproducers of Hindu families. Since coming to power in 2014, Prime Minister Modi has reasserted India's presence on the global stage and strengthened its military power (Stephens 2014; Bloomberg News 2017). This is the archaic modernity that is to bring glory to Hinduism and India.

Development Nationalism

A second example of masculinist rhetoric invoking science, development, and nationalism is evident in the rejoicing that met the Supreme Court's decision in 2000 to allow construction of the Narmada dam. Just a few weeks later, on October 30, L. K. Advani, influential nationalist leader and the government's home minister, remote control in hand, poured a ton of concrete to resume the dam's construction. The construction had been put on hold by antidam activists, who challenged the project in the Supreme Court—struggles in which women were key players. When asked about what he considered the great triumphs of the BJP government, Advani named three: the Pokhran tests in 1998, the Kargil war with Pakistan in 1999, and the Supreme Court verdict resuming the Narmada dam construction in 2000 (Rawat 2000). He saw the last as a victory of the development process and a triumph of development nationalism. The nuclear bomb, a military war, and a big dam are arguably the triumphs of science, technology, and development, shaping the enduring relationships between science, masculinity, and violence (C. Cohn 1987; Nandy 1988). In the speech, Advani severely criticized the protesters who had spent years fighting the megadam project and its anticipated displacement of millions of people as well as its ecological consequences. He questioned their patriotic credentials and wondered if they were being funded by outside sources. Thus, development and the construction of the dam are positioned as nationalist projects while the critique becomes a foreign, un-Indian response.

In the same speech, Advani portrayed the dam as a "victory of development nationalism," highlighting the proposition that religious nationalism is also developmental nationalism.[12] The recent events in India are a strong reminder of the primacy of colonialism within postcolonial, independent India. Despite the rhetoric of decolonization, it is Western science that extends its hegemonic hand to stabilize the Hindu nation. The legacy of Western science lives on as the reason of the state. The return to a "Hindu" past and a call for the revival of the ancient scriptures, the Vedas and Puranas, is an attempt not to decolonize Indian but to reinstate Hindu culture and history as the hub in which the scientific progress of the future is anticipated. There is no epistemological critique of Western science but instead an embrace of it—whereby an exaltation of Western science is simultaneously an exaltation of the scientific Vedas and the Vedic sciences, an exaltation of development and an exaltation of Hinduism.

Early signs of the kind of "development" projects that the new BJP government would undertake suggest that the rhetoric of Indian pride cloaks visions of neoliberal logic to include less state-driven, more privatized models of development. The strident politics of nativism and Hindu nationalism that were the hallmark of Modi's tenure as chief minister in Gujarat (Mehta 2010) portend similar plans for the nation. Soon after coming to power, as a number of news sources had predicted, the Modi government dismantled the Planning Commission, a powerful force in India's national planning since the nation's independence (Pannu 2014; *Hindustan Times* 2014.) Following his campaign proclamation that "Government has no business to do business," early signs suggested that efforts were afoot to privatize many of India's services, such as railways, airlines, mining, and banking (Mantri 2014). In June 2014, a leaked Indian Intelligence Bureau report suggested that the BJP was reinvoking arguments blaming foreign-funded environmental organizations, working against nuclear power, coal mining, and large-scale infrastructure projects, as being responsible for the significant slowdown of India's economy (Ramachandra 2014). Thus, the report argued, foreign influences needed to be controlled to protect India's "national economic security."

SCIENCE, TECHNOLOGY, AND DEVELOPMENT

While religious traditions and practices have always been an important part of India, I want to argue that science has also been deeply inscribed. Although religious nationalists have available a rich set of postcolonial critiques of Western science and development, these have been ignored, and we now have a seamless continuation of previous governments' policies of industrial and scientific development. Far from rejecting Western science, medicine, and technology as one might expect, Hindu nationalists instead embrace it. Nandy and his collaborators suggest that

> Hindu nationalism not only accepted modern science and technology and their Baconian social philosophy, it also developed a totally uncritical attitude towards any western knowledge system that seems to contribute to the development and sustenance of state power and which promised to homogenize the Indian population. There is no critique of modern science and technology in Hindutva, except for a vague commitment to some selected indigenous systems that are relatively more Brahmanic and happen to be peripheral to the pursuit of power. (Nandy, Trivedy, and Yagnik 1996: 62)

When religious nationalists invoke the Vedas or other ancient scriptures in the name of Hindu pride, their vision does not supplant but instead melds with Western science, appropriating Western science within the rubric of Vedic sciences. Rather than claiming a separate and different past and future for Indian knowledge, they embrace Western science *as* Indian knowledge. For example, some argue that many modern scientific developments were discovered or anticipated in the Vedas. Others use the ancient scriptures as a source of pride in the ancient development of literatures, philosophies, and scientific knowledge in ancient India (see, for example, Vivekananda 1992). I want to illustrate this with a few examples of the marketing of Vaastushastra, the ancient Indian material science.

Vaastushastra, commonly seen as the "science of architecture" (Babu 1998), is believed to have been developed four to five thousand years ago in Vedic Indian culture and codified in the Atharva Vedas. *Vaastu* is derived from the word *vas*, which means "to live." Vaastushastra teaches us about "living life in accordance with both desire and actuality" (Chawla 1997: 1). Like the Chinese feng shui popular in the United States, Vaastushastra has become immensely popular in India. While feng shui and Vaastushastra have similarities in positing positive and negative forces, there are also differences in the principles that guide their arrangement of space. My interest in the topic came from watching relatives and friends in India remodeling their houses using Vaastu principles, often at great financial and physical hardship. One rich family with an immense house cooked in the living room for eight months on a tiny electric stove while the rest of the house was closed off for Vaastu-based renovations. Another family bought a newly constructed apartment and then took it apart, down to the tiles on the floor, to renovate it in accordance with Vaastu principles. Living rooms were converted into bedrooms, bathrooms into *puja* (prayer) rooms, bedrooms into kitchens. While individuals undoubtedly renovate homes for many reasons, and while there is a long tradition of religious ceremonies to bless new houses, the recent trend of renovations in order to create Vaastu-friendly homes is an entirely new phenomenon in urban India. I wondered what propelled these individuals to refashion their homes despite the hardship it caused. This rage for Vaastu shapes not only private homes but the public sphere as well. There are now many firms and architects who construct houses, government buildings, and office buildings using these principles. Whether they believe in it or not, architects develop an expertise in Vaastu to satisfy their clients' wishes. The principles of Vaastu, purportedly described in classical texts, are expected to work in any person's house, irrespective of religion or place. The *Handbook of Vastu* states, "Vaastu's concern is not only material property, but also mental peace and happiness and harmony in the family, office etc." (Babu 1998: 3).

The philosophy behind the practice seemed to me at one level practical and at another level ethereal. Vaastushastra considers the interplay of the five elements—earth, water, air, fire, and space—and maintains equilibrium among them. These elements are believed to influence, guide, and change the living styles of not only human beings but also all living beings on earth. The philosophy is ethereal because the literature does not articulate exactly how these elements might bring about health, wealth, and wisdom.

Over the last two decades, this practice has caught on within middle- and upper-class India. In many ways, one should understand the rise of Vaastushastra as a competitive nationalism with China. The Vaastu revival took off in the same period as a revival of feng shui; hawkers were selling feng shui books in the street, at least in northern India.[13] Exploring the phenomenon, I discovered popular books and websites that teach you in a straightforward, detailed, and accessible way the various steps to building or creating a house with Vaastu principles. There are also newspaper and magazine advertisements where experts offer their services. Ultimately I found websites promoting and explaining Vaastushastra to be the most useful and engaging. These sites combined the best qualities of the other two sources—like the advertisements, they were concise and short and engaging, and like the books, they gave more information and background. I used these websites to analyze the claims and practices of Vaastushastra experts.[14]

Consider Mr. Manoj Kumar, who owns the vastushastra.com site. He describes himself as

> an electrical engineer, who has worked with engineering and marketing companies. These experiences gave him the opportunity to observe the functional success of various types of units. In time, he began to assess these premises from Vaastu's viewpoint. His study has included over 3500 industrial, commercial and residential premises. . . . This formed the basis of "cause and effect" relationship in today's context, involving Vaastu principles. Synergizing his engineering skills and four years association with reputed architects to understand the logistics of modern architecture, he is today recognized as Vaastu Consultant of repute.

While the description of his training exclusively heralds his scientific skills—he is an engineer and has worked with many architects—his web page begins with greater claims: titled "Worries, Woes, Wannabes, Winners and VAASTU," the page lists the conditions that Vaastu can help:

If you are disappointed with the ways things are turning out . . .
If you are depressed about the idiosyncrasies of life . . .
If your career graph is plummeting . . .
If your profits are spiraling, but only downwards . . .
If your contemporaries repeatedly hog kudos while you do the lion's share of work . . .
If you find walking on the cutting edge of technology too sharp a going despite your up-to-date training . . .
If your productivity is sub-optimal, and you score only double bogeys in corporate performance when your actual handicap should be under par . . .
If your marriage was made in heaven, but is functional in hell . . .
If your child is the apple of your eye, but the strictest quality standards can't keep the worms out . . .
If you just don't have that "Feel Good" aura about you . . .

The site offers a long list of problems the expert claims Vaastu can help alleviate. In the section "On the Subject of Vaastu," the site further explains:

Vaastu is a complete understanding of direction, geography, topography, environment and physics. It is a study that dictates the form size and orientation of a building, in relation to the plot, soil, surroundings, and the personality of the owner/dweller. There is no room for rituals and superstitions. . . . Impulsive planning and unorganised architectural methods, have led to the primary malady mankind faces today—DISHARMONY. Today more and more architects are turning to Vaastu shastra, to undo this. In short, an ecoimbalance is prevented by synchronising all the Vaastu elements. (emphasis added)

While the site's landing page is an advertisement with claims of improving everything in one's life—happiness, health, wealth, career plans, and marriage—the rest of the description consistently underscores how non-superstitious and nonritualistic the practice is. It does not invoke God or religion as a way to promise health, wealth, and happiness. Instead, it is the promise of science.

Ultimately, the consultant directly confronts the scientific evidence. In a section titled "The Truth About Vaastu," he elaborates:

> There is nothing metaphysical about Vaastu. It is just that the hidden harmony of Nature and the environs around have been defied. Mr. Manoj Kumar in his practice, does not attempt to compare the incomparable or even try to explain it on the basis of magnetic waves, ultra-violet rays, cosmic rays and so forth, as these are but the tips of unknown, unexplored icebergs. The truth is that it works and probably the lowest common denominator of any such venture is its repetitive success, as many conglomerates and inhabitants of Vaastu-designed premises will vouch for. Maybe somewhere, sometime in the future, there will be an explanation for the mechanics of Vaastu, but to dissect it today is a futile exercise.

Some practitioners cite scriptures and ancient records or testimonials from kings, commoners, artists, saints, and traders—all of whom have reaped great benefits from the practice. In this last section, Kumar admits that he cannot provide "scientific" evidence to support his case but suggests that it will one day be found because Vaastu "works" and can be reproduced, and those are grounds of profound proof.

A blogger calling himself Vaastushastra Avanindra (2009) takes up the scientific nature of Vaastushastra, arguing that "Like any other science, Vaastu is considered rational (based on cause and effect), practical, normative (codified and governed by principle), utilitarian and universal." Others further consolidate the scientific basis of Vaastushastra. A prevalent quote online dismisses the religious overtones by arguing,

> Vastu Shastra is a science—it has nothing to do with religion. It is true that over the years, Vastu Shastra has imbibed religious overtones, but that does not seem to have been the original idea. The religious implications were probably inculcated by the proponents

of the science, when they realized that the society of that time was a God fearing one, which would not accept norms that seemed like rituals performed to appease the God.

The website Vaastu International (n.d.) takes the offensive and argues that Western science is much too young to be able to evaluate or appreciate the ancient Vaastushastra:

> The key issue here is, can the modern western science developed over last 300 years can [sic] be used as the only yardstick for assessing super sciences like Vaastushastra formulated over 5000 years ago? A time has come when a broader perspective is needed to study the ancient Indian sciences including Vaastushastra. Pythagorus theorem in geometry was perhaps rediscovered in western world. The principle was successfully applied some 5000 years ago in Indian Vaastushastra. Though shrouded in words like cosmic energy, loading, slopes, cosmic harmony etc., balance between matter and energy is the essential theme of Vaastushastra. Modern version of energy-matter equivalence is found in Einstein's famous theorem $E=mc^2$.

What kind of work does Vaastu do? How and why has it captured India's imagination? After all, different belief systems have always been rampant. Auspicious times have been calculated for jobs, exams, and travel. Indians in all walks of life have consulted astrologers, numerologists, palmists, and so on. Isn't Vaastushastra an extension of those beliefs, like the priest who blesses the house once it is built? At some level I do believe that it is an extension of a belief system that has long existed in India, but the practitioners of Vaastushastra seem to be a departure. They are not priests, astrologers, or individuals claiming divine inspiration, not individuals on street corners or in the recesses of dark, dilapidated homes. Instead, they are engineers and architects, practitioners trained in the sciences (see, for example, Surendran n.d.). The language promoting Vaastushastra is often scientized, such as an architectural firm's promotion comparing Vaastu to "a vitamin pill; it supports you when your starts are weak and gives you a boost when they are strong" (Ray 1998). I have heard anecdotally that many architectural firms and architects have had to learn about Vaastushastra because of a demand from their clients. For example, one architect is quoted as saying, "Whether architects may believe it or not, Vaastu makes a serious impact on the life of the occupant. Designing a structure against the Vaastu rules can and does cause infinite disturbances in the life of the occupants. Just as in cricket, teams prepare favorable pitches to suit them, one can use Vaastu to prepare a favorable pitch for our lives" (*Vaastuyogam* 2012).

Reflecting the broader pattern of the influence of religious nationalism and other aspects of archaic modernity in present-day India, Vaastushastra is literally and figuratively shaping both the home and the world, both public and private spaces. It is especially marketed to women for promoting their good health along with beauty and home decoration tips.[15] Vaastu is directed at women to help the "gracious lady of the house" "[make] it a home," and to help women deliver healthy babies.[16] Numerous websites aimed at Indian women contain instructions and information on rectifying architectural errors (Fernandes 2015). By removing the obstacles of bad luck and spirits through Vaastushastra, women are encouraged to take good care of their families. Some practitioners argue that Vaastushastra principles themselves are gendered, with the angular placing of plots having differential impacts on men and women. Certain configurations are argued to make male members of the family into rogues, turn them to bad habits, or lead them to earn money in unrighteous ways, while other designs can gain women respect in all spheres, cause mental instability, or cause them to be troubled by court litigation![17] Many architectural firms actively advertise Vaastu-informed designs and renovations in newspapers, in women's magazines, and on websites.

Vaastushastra has also shaped public space by its visible presence in the public life of Indian politicians. During the inaugural ceremony of the prime minister, Atal Bihari Vajpayee of the BJP, in 1996, eighteen scholars of Vaastushastra from all over India were consulted about the presence of a lone neem tree in the compound of the main entrance of the BJP headquarters. BJP sources reveal, "Since it was not possible to cut the decade-old tree within an hour, we decided to close the main gate and requested visitors to use the other one. We had no time to get permission from the Union ministry of environment and the directorate of estate to cut down the tree" (Bhosle 1998). N. T. Rama Rao, former chief minister of Andhra

Pradesh, refused to function from the state secretariat in Hyderabad until changes were made to correct its Vaastu. Tamil Nadu governor M. Channa Reddy orders changes every time he moves to a new official residence. B. N. Reddy, a former MP and architect, exhorts state governments to correct the Vaastu of "sick" industries to turn them around (Chopra 1995). Vaastu played an important role in the launch of the Bangalore metro (NDTV 2011). In the recently formed state of Telangana, in 2015 the chief minister hired his trusted vaastu consultant as an "advisor on architecture" to the government (Shivashankar 2015). All these renovations have been completed at the expense of taxpayers.

During the 2014 elections, a newspaper describing one of candidate Modi's appearances wrote, "A vaastu compliant stage that is 100-feet long, astrologers propitiating the gods and invitations sent through social networking sites—the BJP has pulled out all the stops to make its prime ministerial candidate Narendra Modi's maiden rally in Uttar Pradesh at Kanpur October 19 a success" (*Hindu* 2013b). Courting voters in Manipur, in India's Northeast, Modi invoked vaastu in arguing that "Even vaastushastra says that the north-eastern part of a house should be given maximum care. India will prosper only when the north-east will" (NDTV 2014).

Vaastu building practices continue to be highly sought after in public and private constructions, thus literally and figuratively shaping the public and private spheres. Vaastushastra epitomizes archaic modernity, adding "soul" to the architectural sciences, where tradition has infused science with depth, meaning, and history. This is analogous to the nuclear bomb touted to add "soul" to India, where science has infused pride, power, and strength to a once colonized nation. Science, technology, religion, and capitalism have brought Vaastushastra into the market economy of contemporary India, and the system has emerged as another site of frenzied consumerism and a middle- and upper-class status symbol in urban India today.

My use of Vaastushastra as a case study of an archaic modernity is not to demonstrate that there has been little inquiry on whether Vaastushastra is scientific or not, or religious or not, or whether the Vedas do indeed contain all that modern practitioners claim. Instead, I use the case study to illustrate its marketing and consumption as the best merger of the ancient and the modern—a science with soul and tradition. Vaastushastra is a symptom of a visible shift in the social fabric of India, a vision of India as Hindu and a nation celebrating both its ancient wisdom and its new technological breakthroughs.

What we see in the embrace of Vaastu is a facile embrace of dubious claims to a glorious Hindu past, and a celebration of Vedic science and religion. It is frightening to see this Hindu science emerging from nationalism. It is a science that on the one hand purports to be anticolonial, a culturally situated science decolonizing India by unearthing old cultural practices that were eroded by colonialism (Nanda 1997, 1998). Yet in reality, nationalists are seeking to create an India that is a Hindu nation. By finding Western scientific innovations anticipated in the Vedic sciences, the nationalists give India's past an aura of Hindu supremacy. Therefore, in order to look to future progress, they say we must delve into India's glorious Hindu past. This archaic modernity will take us away from neither the problems of religion and science nor those of tradition and modernity. We are deeply implicated in them all. Ultimately, the archaic modernity proves to be much too facile a vision, securing the roots and privileges of the upper-class Brahminic elite. Nothing is threatened except the rights of minorities and women.

IMAGINING INDIA

Archaic modernities invite new interventions in our theories and practices, new alternatives that wrestle with oxymoronic imagery. The challenge is in creating new frameworks in which to think about science and religion. Most responses to these recent developments have tended to glorify prescientific utopias and revive our dreams of a glorious history and our nostalgia for the simple days of yesteryear, a world bereft of scientific and technological innovations, where humanity and technology don't begin to fuse dangerously. Alternately, secular activists have invoked a defense of science, scientific objectivity, and rationality. These critics have historically fought and continue to fight religious nationalism with the rhetoric of science and scientific nationalism. For them, science is our only savior from the superstition and irrationality of religion. At the heart of many of these critiques is the juxtaposing of science and religion as oppositional and mutually exclusive practices.

In this archaic modernity we get to imagine only within the bounded myth of a Hindu India, with a rewritten past, a constructed present, and an impoverished vision of a future. However, for centuries India has been a land with multiple religions and the birthplace of several new ones. These centuries of diversity cannot be erased for a mythical version of Hindu India. The project of decolonization in this age of globalization means taking the project of rethinking "development" and its deployment in the archaic and the modern, the religious and the scientific. We can imagine India in all its rich traditions—Western science, alternate sciences, many religious philosophies and traditions, and indigenous knowledge systems—to re-envision the progressive possibilities of science and religion. We can imagine a new future in an unbounded imagination beyond the worlds of the archaic and the modern.

AVATAR #2

The Story of Amudha

Amudha flowed over Uruvam's cuboidal black body, taking in its radiating worries. The information transformed its liquescent form. Amudha was constituted entirely of liquid—various chemicals and compounds swept through its body at dizzying speeds and in infinite directions. A stream of green flowed down, a flash of pink erupted upward, purple dots swam across the swirling blue, black jets shot through the yellow valleys. The liquids flowed around a central pulsating shaft that appeared to miraculously hold the dulcet flows of its dynamic liquiform. Amudha was always concocting new possibilities. Might some chemical developments on Kari/Earth be reversed? What should it do?

Amudha had always felt a kinship with Uruvam. It was not accidental that earthlings understood the basis of life as chemical. After all, despite Uruvam's efforts, for earthlings all forms were reduced to chemistry. The advent of biochemistry was a day of triumph for Amudha, the node, or Kanu, of Elixirs on Avatara Lokam. But for Amudha the beauty of earthly chemistry was not just in the exquisite structures it had created—the beauty of the benzene ring, the exquisite and varied bonds that connected molecules, the complexity of dodecahedrane, the elegance of the double and triple helices, the infinite and unique diversity of snowflakes as they slowly floated down to Kari. Chemistry at the subatomic level merged into the ephemeral, contextual, and philosophical. Amudha was infinitely pleased with its invention of a playful and clever theater of chemistry for Kari. Embodying Kankavar chemistry, Amudha's molecules were alive, ready to shape-shift, morph, catalyze, and forever recombine into new and interesting compounds. But most humans insisted on seeing matter as inert and dead, fueling a predominantly inanimate view of the world. Many earthlings were puzzled by the quixotic and dreamy playfulness of quantum phenomena. More puzzling results were in store, exposing the Kankavars' cosmic spirit of wonder and mystery of the unknown. Perhaps they would come to understand the animacy and contingency of life?

Unlike Uruvam, Amudha was more pleased with the unfolding of chemistry and chemicals, but Amudha empathized with Uruvam when it came to the larger plans it had for chemistry. Chemistry was not just about the beauty of chemicals but also about what such chemicals enabled. When Uruvam inaugurated a panoply of morphological forms on Kari, Amudha collaborated to make their biochemistries a source of new invention. Could they create organisms that could learn and morph? If skins were porous, might not the chemistry outside permeate the body for new kinds of morphologies? Might chemical forms and organic forms animate each other in a cosmic play? How about the beauty of blemishes and tumors—how cool if organisms could enhance their forms with creative structures around! In an act of exuberance they added warts, moles, growths, lush hair and bald patches, curling nails, and corns. Maybe organisms could nurture and grow these structures into interesting shapes. Maybe, like barbers, skin sculptors would emerge, and everyone would go to them monthly for new ideas. But Amudha had seriously underestimated how seriously humans had mismanaged the environment. In small concentrations, these chemicals Amudha had introduced so playfully would have indeed been a source of gleeful and benign variation. But much to its chagrin, human mismanagement of the environment had resulted in high and lethal concentrations of some chemicals in certain areas, especially in poor countries, regions, and neighborhoods. The chemicals were not easily removed. Large tumors emerged that were soon linked with disfigurement, pathology, and illness. Even benign tumors were deemed ugly and grotesque. Rather than nurturing their beauty, *Homo sapiens* brutally cut into growths and their creations, excising them from the body as gross and misshapen.

Difference, Amudha reflected, was something *Homo sapiens* never learned to embrace. While holding on to some idyllic picture of beauty, they cut and butchered elaborate morphologies into some ideal they had developed. Plastic surgery was invented. Lush and luxuriant bodies were starved, extravagant bodies cut and dissected, and the delicate, dainty body puffed up into monstrous creatures, even though Amudha and Uruvam had ensured that the great Charles Darwin would specifically name randomness, variation, and change as the key ingredients of a bountiful and enduring life on Kari. Evolution is not progress, Darwin had said— nothing can get "better"; all life is beautiful. Change is curiosity, opening the possibilities of new inventions.

While some *Homo sapiens* understood these principles in small ways, they never achieved wide popularity. Increasingly, as the earthlings' avaricious oligarchs colonized the world to crown themselves the lords of the planet, they began polluting the planet with wild abandon. Surely the change-averse species would learn to respect the planet? Alas not, Amudha mused! Organisms sprouted new limbs, lush new growths, new forms of genitalia, and a veritable cornucopia of morphological possibilities. Amudha and Uruvam found some of these new forms beautiful and marvelous. Frogs, fish, mammals, and insects embraced these and wandered Kari displaying their wondrous gifts. Magical multiple limbs, lush asymmetrical growths in wondrous shapes, a plethora of sensory organs to see, hear, small, taste, and feel with. But alas, *Homo sapiens* once again deemed these morphologies liabilities, disabilities, and monstrosities, and soon purity brigades marched the streets demanding extermination. Surely *Homo sapiens* would blame the oligarchs and work to root them out? But what stunned Amudha was that the oligarchs reigned supreme. No one questioned their right to poison Kari. The oligarchs prattled on about growing the economy, producing jobs, building wealth. They should be worshipped, they claimed, not reviled. But of course, the oligarchs did not live in the areas they polluted. They built their own air purifying rooms, distilled their own bottled waters, and grew their own pure food, all the while deeming the toxic chemicals perfectly acceptable for others. The Kankavars were stunned when the oligarchs blamed chemically wrought changes on the victims, portraying them as biologically inferior: they have inferior genes, they are lazy, they are weak, they don't work hard, they are not hardy. Despite Uruvam and Amudha's repeated attempts to curb it, pollution grew unchecked, while a celebration of diversity was increasingly curtailed. The avatars could never get humans to embrace random chance and to play with their creations and inventions. At every turn, the powerful oligarchs controlled Kari. *Homo sapiens* increasingly looked backward to what was rather than forward to what could be. Amudha knew that this could be the end—any species that looked to the past and not the future was surely doomed.

CHAPTER TWO

Colonial Legacies, Postcolonial Biologies

The Queer Politics of (Un)Natural Sex

I did not
        come into being
               a full-grown lesbian
with a knowledge of English,
          a trained brain
               and sexual politics
inscribed upon it.

               SUNITI NAMJOSHI, *Flesh and Paper*

That it really began in the days when the Love Laws were made. The laws that lay down who should be loved, and how.
    And how much.

               ARUNDHATI ROY, *The God of Small Things*

GROWING UP AS A BIOLOGY NERD, I SAW THE "SEXUAL" AS A critical site where the biological and the social easily cohered. I could hold the "naturalness" of plant and animal heterosexuality in my school textbooks and yet recognize that the sexual was socially determined. I could see that foreign films flaunted the bare bodies of their heroines while their Indian counterparts could only suggest the same. As I was working on the various cases of this book, the sexual repeatedly emerged as

76

an important site that revealed the incommensurability of Western and Indian thought. The thigmotropic tendrils of history have braided thick and dense knots around ideas of the sexual, bringing together the diverse genealogies of East and West, science and religion, natural and unnatural, native and foreign, nationalist and antinationalist, pure and polluted, local and global, and the pious and the profane.

In exploring the sexual in India, ideas of natural and unnatural sex shape colonial and postcolonial Indian biopolitics and bionationalism. *Natural* and *unnatural* are held in tension with each other around normative understandings of the sexual. As Sophia Roosth (2017: 67) argues, "'unnatural' blurs the categorical and the normative—it refers simultaneously to that which is counter to nature and to that which is against the (moral) natural order something strange or out of the ordinary. In this regard it bumps up against a similarly charged word: 'queer.'" In queer studies, *queer* has emerged as a capacious term, used as a noun, adjective, and verb; for references to sexuality, gender identity, community; and as a pejorative or a proud affirmation (Blackmore 2011). Here, I use the verb *queer* primarily as method; queering embodies an analytic, forcing us to question how one engages with the world, how one interrogates its foundations and its ideas.[1] In outlining a queer feminist science studies, Cyd Cipolla, Kristina Gupta, David Rubin, and Angela Willey (2016: 19) argue that queering as method signals "openness to the unexpected, the uncertain, and the unknowable" as its defining feature. To queer is to "make strange"; it is an undoing or destabilizing of sexuality and the sexual. It forces us to examine the foundations of knowledge, to interrogate the assumptions of theories, to question what has been deemed natural and normative, and to explore the margins of knowledge for the unexplored and invisible. *Queer* as a verb also "unsettles assumptions about sexed and sexual being and doing" (Spargo 1999: 40). I use *queer* to signal the methodological impetus to "open up" the term *unnatural sex*, to explore what emerges when we bring feminist science studies alongside queer theory and postcolonial studies. I trace the unexpected and queer travels of ideas of the sexual in the colonial and postcolonial state. How do the categories of "natural" and "unnatural" sex emerge in plant and human biology, in British colonial ideologies, and in Hindu nationalist discourses? While the British deemed homosexuality and its oriental colonial subjects as sites of sexual otherness, Indian nationalists, in turn, relegated homosexual, indigenous, and Muslim Indians as their sexual "others." These various strands, I argue, are linked. A recent case that helps illuminate these thigmotropic formations is the debates surrounding Section 377 of the Indian Penal Code (IPC 377), the focus of this chapter.

QUEERING SEX: NATURAL
AND UNNATURAL HISTORIES

What is natural sex? When is sex unnatural? This chapter emerged from the recent debates over Section 377 of the Indian Penal Code (IPC), often referred to as the "anti-sodomy" statute. A typical Indian story is undoubtedly a queer one, as our investigations find stories within stories within stories, where science and religion, colonial and postcolonialism tumble into each other, all intertwined into a multidimensional, braided thigmotropic knot. I explore these complexities in five sections. Act 1, "Indian Penal Code 377," outlines the complex history of the origin and implementation of IPC 377. The statute is fundamentally constituted through the key term *unnatural*, a term that becomes the focus in the rest of the essay. Act 2, "Sex and Science: The Judeo Christian Roots of Natural and Unnatural," explores how the history of biology was shaped by Judeo-Christian imaginaries and came to enshrine divinely inspired and sanctified Victorian sexual politics in the reproductive biology of plants, animals, and humans. With this understanding of sex as the cornerstone of the biological sciences, "natural" sex came to undergird scientific and civilizational logics ever since. In act 3, "Genealogies of the Sexual in India," I explore how the emerging sexual politics of Indian and Hindu nationalism shaped and were shaped by British colonial views on natural and unnatural sex and sexuality. In act 4, "Religion, Science, and the Discourse of 'Natural Sex,'" I explore how the science and religion of the "natural" intersect through two sites: the eugenic roots of Indian nationalism, and key arguments in recent legal challenges to IPC 377. Finally, in act 5, "Rescripting the Sexual: Cultivating Postcolonial Imaginaries," I discuss Hinduism's multiple strands and how this diversity, along with queer social and scientific movements, might interrupt colonial sexual scripts to open up new sexual imaginaries and possibilities.

I want to highlight two terms in particular that I use throughout this chapter. First, the term *queer* for identity groups is a decidedly insufficient translation for the diverse sexualities India has engendered (Bacchetta 1999; Puri 2012; Narrain and Bhan 2005; Dave 2012), but no term overcomes the incommensurability of language and social ontologies between India and the West. There are many well-established sexual identities in India: to name just a few, *kinnar, aravani, hijra, kothi, jagappa*. These identity groups have long been historically embedded in local cultures (although often marginalized) and have been woven into local identitarian understandings (Kannabiran 2015). They are not nonnormative in the Western sense of the term. In the Western literature, other terms include *same-sex sexuality, gay, lesbian, bisexual, trans*, and *men who have sex with men* (MSM). None of these terms are universally accepted (Puri 2012), and they do not always translate to other regions of the world. Similarly, the term *queer* doesn't translate easily into Indian contexts (Bacchetta 2013). In short, there is no ideal, universally accepted term that acknowledges the complexities of sexual identities.

Second, I use the terms *Indian nationalism, Hindu-centric*, and *Hindu nationalism*. Broadly, *Indian nationalism* refers to mainstream nationalism that emerged during the Indian freedom struggle against the British. It is characterized as secular, imagining India as a multireligious, secular, democratic republic. In contrast, *Hindu nationalism* posits South Asia as once the land of the Vedas and the Vedic civilization, and therefore India as a nation that should be Bharat (after Emperor Bharata of the Hindu epic *Mahabharata*) or Hindustan (land of the Hindus). While *Indian nationalism* is imagined as more inclusive, it too draws heavily from Hinduism. Much of Indian nationalism is deeply Hindu-centric—drawing on India's Hindu past to posit a great ancient civilization. The boundaries between Indian nationalism, Hindu centric nationalism, and Hindu nationalism remain porous and murky.

ACT 1. INDIAN PENAL CODE 377

Histories and analyses of IPC 377 are by now well-researched territory, and many scholars have written eloquently about the law's colonial roots and the postcolonial politics that have kept such laws on the books

(Bhaskaran 2002, 2004; Narrain 2004; Kapur 2005, 2009; Puri 2012, 2016; Shah 2015; Shandilya 2016). What has been less explored is the term *unnatural*. Coming from the field of feminist science studies, I know that the term has deep roots in queer and feminist studies and the biological sciences (Cronon 1996; Haraway 1997, 2000; Willey 2016).

IPC 377 is one of 511 criminal statutes that compose the Indian Penal Code. Drafted by Lord Macaulay in 1860, while India was still under colonial rule, it resides in the subchapter "Of Unnatural Offences" and criminalizes all sex other than heterosexual penile-vaginal sex between a man and a woman (*Naz Foundation v. Government of NCT Delhi* 2009). It reads:

> 377. UNNATURAL OFFENCES—Whoever voluntarily has carnal intercourse against the order of nature with any man, woman or animal, shall be punished with imprisonment for life, or with imprisonment of either description for a term which may extend to ten years, and shall also be liable to fine.
>
> *Explanation*—Penetration is sufficient to constitute the carnal intercourse necessary to the offence described in this section.
> (Indian Penal Code 1860)

When IPC 377 was introduced as law, it brought into the legal code prohibitions that had no equivalence in India. It was thus a specifically colonial invention (Puri 2016). Scholarly consensus is that it inaugurated the category of the "homosexual" into Indian legal history, its focus being not the transgressive act but the transgressive person: "the consenting homosexual" and the "habitual sodomite" (Puri 2016). These terms, emerging from the ideologies of the late nineteenth century, introduce an innate and biological category of humans, ones "predisposed to crime through habit or heritage" (Puri 2016: 73). Kath Weston (2008) shows how the prisons in the Andaman and Nicobar Islands become the site for some Indian prisoners convicted under IPC 377 during colonial rule. As she argues, "Yet what constituted an unnatural offense was—and is—not at all obvious" (Weston 2008: 218). In a historic ruling, on September 6, 2018, the Supreme Court of India, reversing an earlier Supreme Court decision, decided that the application of Section 377 to consensual homosexual sex between adults was unconstitutional, "irrational, indefensible, and manifestly arbitrary." As we shall see, other acts deemed "unnatural sex" remain illegal under IPC 377.

As in India, the legal code remains in all Asian former British colonies. Of all the colonial powers of Western Europe—Britain, France, Germany, the Netherlands, Portugal, and Spain—only Britain left such a legacy in its colonies (Sanders 2009). Sanders (2009: 8) shows how the roots of the anti-Catholic buggery laws of 1535 now "remain hidden behind the language of the late 18th century Indian Penal Code."

The sexual subject in India emerges from a very different set of biopolitical evolutions than the Foucauldian subject of the West. Foucault ([1976] 1998), in his first volume of *The History of Sexuality*, posits that the growing discourse on sex in the nineteenth century turned sex into a "problem of truth." Sex emerged as something potentially dangerous and disruptive and became a site of "knowledge" rather than morality. Foucault distinguishes between the West, which dealt with sex through a *scientia sexualis* (science of sexuality), and countries in the East (Rome, China, Japan, India, and the Arabic-Muslim world), where knowledge of sex was private and passed on by the *ars erotica* (sex as sensual pleasure).[2] While he subsequently expressed doubt about the difference between Western and Eastern discourse on desire, he never entirely disowned the distinction in his work (Rocha 2011). Foucault's work on *scientia sexualis* helps us understand how the development and regulation of sex as an object of science produced a "uniform" truth of sex through science. While the distinction between *scientia sexualis* and *ars erotica* is indeed not clearcut in the West, India also defies any easy distinction between sexual science and sexual pleasure—both are deeply interwoven in the histories and philosophies of science and religion in India. Even while elements of sexual science were explored and imposed on India by the British, contestations by Indian nationalists impose counter theories of sexuality.

Jyoti Puri chronicles the paradoxes of the implementation of IPC 377. As Puri (2016: 59) writes: "case law on Section 377 does not so easily yield. Rather than figuring the sodomite, it turns out to be a collection of sexual offenses, including consensual and nonconsensual sex between adult men, sexual assault on children and women by adult men, anal and oral sex coerced by men from their wives, and bestiality." As it turns out, the vast majority of IPC 377 case law involves sexual violence against children.

Despite its sporadic formal use, the antisodomy statute became the centerpiece of queer activism in India. The law's biggest impact was its use as a weapon of policy abuse, particularly by the police: "detaining and questioning, extortion, harassment, forced sex, payment of hush money." While rape laws are written to protect the victim, IPC 377, in contrast, is meant to punish. By linking "carnal intercourse against the order of nature," the law creates a tautology between unnatural sexual practices and criminality (Puri 2016).

The success of recent efforts to repeal IPC 377 are part of larger social transformations of sexuality in India. Women's movements, the public health crises with the rise of HIV/AIDS, and the liberalization of the economy and the media worked together to reconstitute debates on Indian sexuality (Kapur 2005; Jones 2014; Shah 2015), all focused on the importance of repealing 377. The growing public debate on sexuality emerged for several reasons—the transnationalization of "LGBTQ" identities, sustained media and policy debates, increasing numbers of people who identify as queer and trans, more spaces to socialize, the rise of gay pride events and protests, growing social media networks, and greater visibility of queer politics (Shah 2015).

In India, 377's most egregious impact had been to condone physical violence and discrimination of same-sex sexualities in the name of "public morality," thus supporting the "intolerance and violence committed in intimate spaces of the family as maintenance of social order" (Puri 2012). The law gained public attention when the police superintendent of Tihar jail, Kiran Bedi, refused to hand out condoms to inmates who risked contracting HIV/AIDS through unprotected anal sex (Shandilya 2016). The activist movement AIDS Bedbhav Virodhi Andolan (ABVA) mounted the first official challenge to 377 in 1992 when it petitioned the Parliament for a repeal of the law. ABVA never received a response (Bakshi 1996). After a failed attempt in the 1990s, the Naz Foundation challenged the law's constitutional validity in 2001. In July 2009, the Delhi High Court, in a historic judgment, overturned the 150-year-old section, thus legalizing sex between consenting adults, affirming that 377 goes against the fundamental rights of human citizens. The central (secular) government could have appealed the decision but chose not to. However, a group of religious fundamentalist activists of Hindu, Muslim, and Christian faiths appealed the high court's decision (Shah 2015). "All religious communities—Muslims, Christians, Hindus—had said that this was unnatural sex," proclaimed Ejaz Maqbook, the lawyer representing the religious groups (Gowen and Lakshmi 2013). The case was heard in 2011, and in December 2013 the Supreme Court of India set aside the 2009 verdict, reinstating IPC 377. In upholding the law, the Supreme Court claimed, "LGBT people constitute a miniscule minority," and therefore revoking 377 is unnecessary (Shah 2015). In February 2016, the Supreme Court heard a curative petition and referred it to a five-judge constitutional bench for a possible "back-to-roots, in-depth hearing" (K. Rajagopal 2016). On January 8, 2018, the Supreme Court agreed to reconsider the constitutional validity of IPC 377 (*Express Web Desk* 2018). On September 6, 2018, the Supreme Court unanimously struck down one of the oldest bans on consensual gay sex (Gettleman, Schultz, and Raj 2018), ruling that gay Indians should be accorded all the protections of the Constitution. They called the original law "indefensible," reversing the earlier ruling by arguing that "majoritarian views and popular views cannot dictate constitutional rights." Justice Indu Malhotra called for a "national apology" to the LGBT community that had lived under stigma for over a century (Dutt 2018).

The decriminalization of "gay sex" was significant. However, the debates that led up to this victory are still illustrative of the complex biopolitics of sex and sexuality in India. Groups opposing IPC 377 coalesced around several questions that shaped the Court's decision. First, was homosexuality "normal" and "natural"? Was the Indian public ready to accept it? Some argued that it was both normal and natural, and that IPC 377 was an "obsolete" law enacted by the British, who had since repealed it in their own nation; India should follow (Vanita 2007). Others argued that homosexuality was both unnatural and not normal on scientific, cultural, medical, and religious grounds. Second, was homosexuality Indian? Those wanting to revoke IPC 377 argued that the law was thoroughly inconsistent with Indian tradition, in which homosexuality is not a crime. Activists quoted from academic work on same-sex love in India, arguing that precolonial India had "generally tolerant traditions" (Vanita and Kidwai 2000: 194). In Indian mythology, for example, the god Ayyappan is said to be born to male gods Shiva and Vishnu. Further, they argue that since IPC 377 was first imposed by the British, it was in fact "a form of cultural imperialism imposed by a colonial power, breaking a longer permissive history in India" and introducing into India a still-flourishing "homophobia of virulent proportions" in the late nineteenth and twentieth centuries (Vanita and Kidwai 2000: 200). They invoked the rich legacies of erotic arts in temples and caves, the *Kamasutra*, and other erotic texts to claim that homosexuality was indeed constitutive of Indian culture (McConnachie 2008; Vanita 2013). Third, claimants argued that IPC 377 should be revoked for public health reasons. Questions of whether homosexuality was normal and natural have been central to public policy and health debates around condom distribution to prevent the spread of AIDS (Kapur 2005). Some argued that IPC 377 was a setback for public health, crippling prevention efforts for HIV/AIDS (Kapur 2005). In sum, Section 377 not only criminalized some sexualities as nonnormative but more importantly *institutionalized* unequal rights and a lack of the protections of citizenship (Puri 2012).

ACT 2. SEX AND SCIENCE: THE JUDEO-CHRISTIAN ROOTS OF NATURAL AND UNNATURAL

> By making the normal the object of a thoroughgoing historical study we simultaneously pursue a pure truth and a sex-radical goal: we upset basic preconceptions. We discover that the heterosexual, the normal, and the natural have a history of changing definitions. Studying the history of the term challenges its power.
>
> JONATHAN KATZ, "The Invention of Heterosexuality"

The "unnatural" nature of sex in IPC 377 (and its original conceptions during British colonial times) can best be understood through a *naturecultural* lens. In the "natural," science and religion come together to distinguish the desirable from the undesirable, the normal from the abnormal, the civilized from the uncivilized, the normative from the deviant, and the pure from the profane. As Katz (1990) argues, the heterosexual, normal, and natural emerge as mutually constitutive biopolitical terms. Here, I want to link the histories of biology and religion in the constitution of the "natural" and the category of "unnatural offenses."

Nature and the "natural order" are themselves concepts deeply imbued with Christian imaginaries. Nature was God's creation, and early Western scientists saw their role as explicating divine wisdom through a study of the natural world (Moreland 1989; Noble 1992). Christian ideologies and morality found themselves undergirding scientific theories and logics. "Magic" and witchcraft, once largely the purview of powerful women, were routed out of early Western societies because they were considered as un-Christian; the result was the rise of Western science and various forms of Christianity as the sole arbiters of power (L. Gordon 2006). In colonial states like India, local knowledges and indigenous and animist traditions were equally reviled and dismissed as backward, unscientific, and archaic (Prakash 1999).

What constitutes natural and unnatural sex and sexuality? Categories of "unnatural" sex traverse social, political, cultural, medical, and biological realms—and the politics of gender and race figure prominently. As queer feminist STS has long taught us, the power of science is in its extensive networks and links to other structures of power. In regulating sexuality, legal codes like IPC 377 define and govern it, social institutions of family and the state enforce what constitutes the "normative" and permissible, and religious leaders codify what is acceptable and not. As Foucault argues, science/biology/medicine came to define the normative and the deviant/unnatural/perverse in the late nineteenth century. IPC 377 demonstrates the power of the "natural," a critical node around which many institutions cohere—religion, science, medicine, and the law.

Nature and natural law was God's will (Moreland 1989). Thus, nature was simultaneously scientific and religious. The natural order that scientists and naturalists outlined became the foundation of modern Eurocentrism, a foundation that was exported to the colonies. Through Eurocentrism, the European eye created order through a "system of nature" that produced European discourse about non-European worlds. Something like the Linnaean classification system was exported so widely that it is "comparable to Christianity," and Linnaeus is an "ambassador of empire." According to Mary Louise Pratt (1992: 28), "The naturalist naturalized the bourgeois European's own global presence and authority." This natural order explicated a particular European sexual and racial geography of life that was then projected onto plants and animals. In turn, the widespread "naturalness" of plant and animal sex and sexuality came to reinscribe European beliefs in the intersecting conceptions of gender, race, and sexuality of Europe as the "normative" or "natural" mode of planetary consciousness. As Pratt (1992: 36) asks, "For what were the slave trade and the plantation system if not massive experiments in social engineering and discipline, the systematization of human life, the standardizing of persons?"

The development of the biology of plant sexuality is a very queer tale. "Hermaphroditic plants 'castrated' by unnatural mothers. Trees and shrubs clothed in 'wedding gowns.' Flowers spread as 'nuptial beds' for a verdant groom and his cherished bride. Are these the memoirs of an eighteenth-century academy of science, or tales from the boudoir?" asks Londa Schiebinger (1993: 11) in her well-regarded book *Nature's Body*, which traces the changing foundational conceptions of sex, gender, sexuality, and race in biology. As Schiebinger reveals, these are categories Carl Linnaeus developed to revolutionize the study of the plant kingdom. Previous to Linnaeus, within medieval cosmology, plants were understood for their primary purpose: as food and medicines. In the eighteenth century, botanists shifted their classification of plants from a more local form of "useful" and hands-on knowledge to a scientific classification that was both abstract and universal. Concomitantly, the "'scientization' of botany coincided with an ardent 'sexualization' of plants" (Schiebinger 1993: 12). Schiebinger argues that the scientific revolution and the revolution in sexuality and gender came together to elevate plant sexuality as a central focus of botany in the eighteenth century and a key to classification.

The "laws of nature" were read through the evolving lens of human "sexual relations." The implicit use of gender to structure botanical taxonomy came together with the explicit use of human sexual metaphors to create a new and innovative classification system for plants. Linnaean taxonomy came to be built from new understandings of human sexual difference and traditional notions of colonial sexual hierarchy, imbued with imported vivid sexual language of humans. The importance attached to sex by both church and state, and their regulation of sexuality, shaped the growing importance of sexuality's biopolitical dimensions.

Linnaeus gave a primacy to plant sexuality that no other naturalist had attempted. As Janet Browne (1989: 597) argues, "to be a Linnaean taxonomist was to believe in the sex life of flowers."[3] Linnaean classification was based on a fundamental difference between the male and female parts of flowers. Linnaeus divided the "vegetable world" into classes based on male reproductive parts. These classes were subdivided into orders based on female parts. Of course, males were always superior to females. Orders were then divided into genera, species, and finally varieties.

Linnaeus based his system on a binary understanding of sexual difference, and the "laws of nature" were read through the evolving lens of human sexual relations, bringing traditional notions of gender hierarchies "whole cloth into science" (Schiebinger 1993). His renowned "Key to the Sexual System" is founded on the *nuptiae plantarum* (the marriage of plants), the idea of their union and implicit heterosexuality, not their sexual dimorphism. As marriage in the seventeenth and eighteenth centuries shifted from arranged to love-based, the flowery language and romance blossomed to animate a florid dictionary for the sexual lives of plants. As Janet Browne (1989: 600) argues:

> It is worth emphasizing here that it was Linnaeus who initiated this personification of the sexual relations of plants and that his more robust followers were merely accepting and extending the practice into English-language works. This use of personification allowed Linnaeus to write of plant sexuality as a "marriage" and the male and female organs as "husbands" and "wives"; he wrote of the petals (corolla) as the "marriage bed"; and he discussed the existence of *monoecious* and *dioecious* plants in terms of one or two different "houses." By coining the words *monoecious* and *dioecious* (derived from the Greek for one or two homes or houses), Linnaeus set up a system of metaphors through which plant sexuality could be made intelligible by being modeled on human society, in much the same way as La Fontaine's moral fables owed their dramatic force and piquancy to their location in the animal world rather than the human.

Linnaean classification in *Systema Naturae* also extended to humans, to fortify the precision of race. His impetus was to isolate and fix race into a constant to create a stable classification system (Niro 2003). While scientific racial types shifted over time, the "logic" of race long endures as perhaps Linnaeus's most far-reaching legacy. The colonial legacy of botany is immense and deep; colonial expansion permeated the globe, and Western systems of nomenclature supplanted local knowledges. Jamaica Kincaid (1999: 122) evocatively notes, "This naming of things is so crucial to possession—a spiritual padlock with the key thrown irretrievably away—that it is a murder, an erasing." Colonial legacies have dismantled and erased a plethora of languages, meaning-making practices, and nomenclatures to usurp multiple cultures of knowing and replace them with a universal, scientific "monoculture of knowledge" (Shiva 1993). The sexuality of plants and humans thus shares an intimate origin story that traces the histories of colonial conquest and rule.

The Linnaean classification system fundamentally transformed European social mores, divinely ordained, into "natural" attributes of an abstract "nature" that was also under divine control. Biologists continued to build on this idea. Enlightenment thinker Erasmus Darwin, for example, worked to show plants as an integral part of animate nature, arguing that all life, including plants, had the same attributes as animals (Browne 1989). For Darwin, like Linnaeus, sexuality emerged as a normal feature of human life, and love emerged as a "natural law."

Linnaean classification ushered in a larger "planetary consciousness" that shaped European colonialism and exported a "natural" order, shaped by Christian morality under the guise of science (Pratt 1992). Thus, on the one hand, Enlightenment thinkers put certain Christian ideas about "natural" and "unnatural" sex on a "scientific" footing, and then social reformers invoked the epistemic power of the same scientific natural and unnatural order to reorder and reorganize social life, especially sexuality, across colonial states (R. Lancaster 2003).

Subsequent naturalists and biologists, including Charles Darwin, helped update the idea of the scientific "natural," continuing to shape our understandings of natural and unnatural sex and sexuality. Questions such as the "nature" of desire, the utility of "sex," and resulting ideas about fitness, normality, and degeneracy shaped scientific studies of the laws of nature. Indeed, theories of evolution imagine not the individual but rather the white heterosexual couple as the pinnacle of evolution (Markowitz 2001). Thus, science developed Christian-inspired parables of "natural law" governing divinely designed plants and animals (R. Lancaster 2003: 107).

If plant and subsequently animal sex was modeled on the primacy of human heterosexuality, the preoccupation with sexual dimorphism and heterosexuality as the only "natural" modes of biology has endured well into the twenty-first century. Plant reproductive biology and Darwinian evolution are exclusively focused on models of the vertical transfer of genetic material—that is, sexual reproduction as the exclusive focus of biology. Jennifer Terry (2000), in her wonderful and sometimes hilarious essay on "unnatural sex," chronicles science's fascination with queer animals. Feminist scholars have also powerfully explored how binary sexual categories and gendered language continue to undergird the heteronormative compulsions of contemporary biological sciences (Warren 1990; Kinsman 2001; Hird 2004; Pollock 2016). The idea of queerness in animals and nonprocreative sexual behaviors and acts continue to confound evolutionary theory that is grounded on the primacy of sexual reproduction. The deep logics of heterosexualism and its procreative imperative continue to undergird ideas of the natural and normative. Nonprocreative organisms are evolutionary "dead ends" by this logic, and yet, as Bagemihl (1999), Terry (2000), Roughgarden (2004), and other scientists demonstrate, they are rampant in the plant and animal worlds. Undoubtedly, a queering of the logic of the biological imperative remains fertile ground for feminist STS

ACT 3. GENEALOGIES OF THE SEXUAL IN INDIA

> The reconstruction of the past implies a clash of stories deeply enmeshed in the discursive construction of present identities. That is why history is so important, because it is part of what we think we are; it is part of our culture.
>
> PETER VAN DER VEER, *Religious Nationalism*

We have so far explored the law criminalizing "unnatural" sex in contemporary India, and how Christianity, the biological sciences, and colonial expansion brought Victorian sexual morality into plant biology and sexual norms. But how did these translate into India? What happened when colonial ideas of the sexual confronted the sexual of India? Jyoti Puri (2002b: 603) begins her wonderful essay on the *Kamasutra* with the words, "I discovered the *Kamasutra* through the eyes of the West. The *Kamasutra* was not an integral part of the lives or the sexual development of adolescents like myself coming of age in India." This has deep resonance for me. Despite the ubiquitous presence of the sexual in India—in worshipping the aniconic representation of Shiva in the phallic *lingam*, in the ubiquitous sexual art in temples and caves, in the histories of South Indian classical dance and music, in the Devadasi traditions of temples (young girls who were "dedicated" to deities or temples for all their lives)—the sexual was little discussed (Coorlawala 2005). Paradoxically, while I missed the sexual overtones in the religious worlds around me, my friends and I were deeply aware of the puritanism of Indian postcolonial society. The best examples were the Bollywood films we copiously consumed. As we joked, the minute the hero came close to the heroine and as their lips approached, a large elephant would walk by! In urban India, we were deeply aware of the sexually repressive culture, which we enjoyed berating, but at the same time we were ill informed of a different set of histories and genealogies of the sexual in India.

As I grew up and explored these histories, I encountered a widespread narrative in secular India: there was a "golden era" in India's precolonial past that was sexually adventurous, permissive, and open to different genders and sexualities. The sexual iconography in temples and caves can be traced to religious and spiritual traditions of eras long past, drawing on the varying practices of *kama*, *tantra*, and *yoga*. This narrative celebrated India as a site of multiplicity and heterogeneity. It argued that a polytheistic religion like Hinduism, which has a multitude of gods and innumerable animist traditions, was bound to spawn a rich and varied sexual culture and fantastical narratives of spiritual and sexual exuberance. Gods in Hindu mythology have vibrant romantic, sensual, and sexual lives. The quintessential stories of Krishna and his incorrigible flirtations have inspired many classical songs and dances. In Bollywood movies, the

Mughal courts were depicted as deeply sensual and sexual sites. The courtesans commanded the stage with their grace and poetry. One journalist captured these sentiments: "[That] this extremely conservative country was once home to the world's first sex treatise and the erotic art on display is perhaps more shocking now than when it was created" (Ramadurai 2015).

This secular narrative celebrated the sexual exploration and permissiveness of India's sexual past via romanticized accounts of a glorious ancient India (Puri 2002a, 2000b). According to this narrative, India has always presented us with pluralistic and promiscuous possibilities. It is only colonial rule, with the puritanism of Victorian sexuality and the excesses of orientalist logic, that erased this past. The result was a divergence between Indian mythological extravagance of the sexual and its more puritanical incarnation in everyday life in postcolonial India. It is a seductive narrative.

Was India's past sexually vibrant? In exploring the sexual in India, orientalism looms large and presents an overdetermined representation of the sexual. As Anne McClintock (1995: 22) has argued, India and other nations became the "porno tropics," where lascivious and libidinal desires forbidden in the West found a home:

> For centuries, the uncertain continents—Africa, the Americas,
> Asia—were figured in European lore as libidinously eroticized. . . .
> Renaissance travelers found an eager and lascivious audience
> for their spicy tales, so that, long before the era of high Victorian
> imperialism, . . . [these continents] had become what can be called
> a porno-tropics for the European imagination—a fantastic magic
> lantern of the mind onto which Europe projected its forbidden
> sexual desires and fears.

Here, a luxurious and copious orientalist vision turned India into a site of the exotic and erotic—the East filled with danger, pleasure, deviance, permissiveness, promiscuity, and perversity (McClintock 1995). The Orient, including India, became the "sexual other" of Western sexual superiority.

Repeatedly in India, one confronts contradictory histories. How do we reconcile the secular narrative of a golden past with the excesses of orientalist inventions? Who is right? As postcolonial scholars have argued,

histories of postcolonial nations are always written in the shadowy machinations of colonialism. In trying to reconcile true and authentic histories,
the past becomes contentious territory. First, we need to account for the
orientalist tendencies of the discipline of history. Trying to decolonize
the historical record has been an immense and ongoing project for postcolonial studies. Second, reading and writing between the "porno-tropics"
of orientalist histories and the erotic figures on temple walls is always a
work of translation. As Puri (2002b: 615) ably argues in her history of the
Kamasutra, "originals" and "translations" are "part of a larger discourse of
making and dispensing unequal histories within the colonial context."
Every account of the past is a translation of sorts (Epstein, Johnson and
Steinberg 2000), and historians of ancient India inevitably transform their
present into the past—whether it is about the relationship between colonial India and colonial Britain or a postcolonial expression of a hybrid
modernity (Puri 2002b). Thus, accessing an "original" or "authentic" history is fraught with the contentious politics of coloniality and postcoloniality. As Anjali Arondekar (2014) reminds us, secular narratives of
sexuality in India inevitably are grounded in a language of loss, and she
warns us about such narratives of loss and recovery. In contrast, her
important work offers and reveals to us the rich and largely unexplored
archives that would suggest a history of the plenitude of sexuality in India.
Perhaps the only solution is to write into the history of modernity the
many contradictions, ironies, and ambivalences that history offers
(Chakrabarty 1992). We should then acknowledge the sexual exuberance
on temple walls and caves, the vibrant sexual forms of art and music, the
porno-tropics of orientalism, the unexplored sexual archives, *and* the
resulting puritanical politics of elite India, locating and understanding
each within the very queer colonial and postcolonial histories of India.

As the British crafted the Orient, including India, as its "sexual other,"
Indian nationalists responded to colonists' orientalist exuberance with
revulsion, indignation, and outrage, rejecting such views as colonial
mythology. The Indian independence movement worked through a "bifurcated" conception of the nation. On the one hand, they acknowledged the
achievements of the British—and India's own inferiority in this respect
(in being colonized)—but they simultaneously presumed India as a site of

inner and cultural superiority (P. Chatterjee 1993). In the emerging nationalist movement, the elites of the Indian nationalist movement compromised with religious orthodoxy, creating a tension between progressive secular nationalists and conservative religious nationalists—a tension sustained in contemporary Indian politics (A. Rajagopal 2000).

The Indian independence movement emerged as a Hindu-centric movement, where claims of the inner and cultural superiority of India came from its Vedic or Hindu past. Hindu nationalists challenged the weak and effeminate politics of nonviolence and advocated an armed response to British colonialism. It was this wing of the movement, a member of the Rashtriya Swayamsevak Sangh (RSS), that assassinated Mahatma Gandhi. The group was subsequently banned. It is telling that sixty-eight years later, a former leader of the RSS is now prime minister of India. The ideologies of the RSS are worth exploring because they are an illustrative thread in explaining the incommensurability of sex and sexuality between India and the West.

In tracing the genealogies of Hindu nationalism on sex and sexuality, Paola Bacchetta (2013) makes certain key points. First, Indian nationalist ideologies of sex and sexuality were shaped by British orientalism and colonial law. Second, the British used caste politics to divide India. They discredited the upper-caste Brahmins (who held symbolic power), coopted the rulers of princely states (Kshatriyas, who had material power), and created a civil service and army with Indian collaborators who helped them rule. The Brahmins were constructed as effeminate, and the Kshatriyas, members of the warrior and princely caste, emerged as the ideal of Hindu masculinity. Third, the British shaped their civilization mission in part around constructing India as a site of the "porno-tropics," where Indians were constructed as sexual others—upper-caste Hindu men as oppressive to women and lower-caste individuals as sexually promiscuous.

While the British—through science, Christianity, and colonial discourse—promoted heterosexuality as the "natural" state of humans, this portrayal did not translate completely in the Indian context. Hindu nationalist organizations condemned homophobia but at the same time celebrated celibacy in their leaders. For example, the RSS, the premier Hindu nationalist organization that emerged during the Indian independence struggle, is a fundamentally homosocial organization. It describes itself purely in masculine terms: "the men born in the land of Bharat" (Golwalkar 1980: 107, 208). It is constructed as a "fraternity of men" and "sons of the soil," whose highest leaders are heterosexual but self-proclaimed celibates. Their organization is characterized by "deep horizontal comradeship." M. S. Golwalkar (1980: 291), a key RSS ideologue and its second supreme leader, described the organization's mission:

> Let us approach every son of this soil with the message of one
> united nationhood and forge them all into a mighty organized
> whole bound with ties of mutual love and discipline. Such an alert,
> organized and invincibly powerful national life alone can hope to
> stand its head erect in the present turmoils of the war-torn world.

The RSS's ideal bond of "love" rests on the Hindu male repression of homosexuality—drawing a distinction between friendship and its distinction with the erotic and sexual. In an all-male organization, homosexuality threatens to disrupt male-to-male homosocial intimacy. Thus, Bacchetta (2013) argues, the operative sexual binary for Hindu nationalist men is not hetero- and homosexuality but rather asexuality versus both homo- and heterosexuality. In this logic, "Hindu nationalist men must avoid all sexual contact in order to remain faithful to the collectivity of Hindu men" (Bacchetta 2013: 149). It should not surprise us that such an organization promotes celibacy in its leadership. The *sarsanghchalak pracharaks* (full-time RSS workers and the self-proclaimed leaders) and *swayamsevaks* (RSS members) who are permanently celibate are the most ideal of Hindu men. It is also significant that Hindu leaders, who are celibate yet virile and strong, are inevitably bigendered, inhabiting male and female characteristics. They may show a mother's love and a father's diligence, be both poet and warrior, or be described as militaristic yet childlike (Bacchetta 2013).

From this genealogy emerges the ideal rank-and-file Hindu nationalist: "a virile, chivalrous warrior along Kshatriya lines; celibate along Brahminical *sannyasin* (wanderer detached from material world) lines, and 'respectful of women'" (Golwalkar 1980: 449, 588). As Alter (1997) argues, sex is worked out in "somatic rather than psychological terms," defined in terms of male physiology and gendered conceptions of good health. This is a quintessentially Indian and Hindu nationalist mode of masculinity, disruptive to a Western scientific and religious normative heterosexuality. In particular, this queer vision contrasts with British masculinity during colonial times and Western masculinity ever since. Western conceptions of normative masculinity are deeply grounded around a "sexual" male libido, around Darwinian sexual excess. Ideal Western males are sexually promiscuous, sowing their seeds far and wide. Sexual prowess is key to normative masculinity in the West but fundamentally undercut in Hindu nationalism. Thus, Western ideas of normative sex and sexuality do not translate completely into the Indian context, and vice versa.

Hindu nationalists translated the edicts and practices of celibacy as a claim to Hindu superiority, with heightened control and discipline of the mind and body. By labeling normative Western sexuality as hypermasculine and hypersexual, Hindu nationalists reclaimed their moral and physical superiority of mind, body, and reason, rendering Hindus as chaste, orderly, heterosexual, and virile (Bacchetta 1999). As Golwalkar (1980: 14–15) characterizes the British: "The insatiable hunger for physical enjoyment does not allow one to stop within one's own national boundaries. On the strength of its state power, the stronger nation tries to subdue and exploit the other in order to swell its own coffers. . . . Moral bonds are all snapped." A "muscular masculinity" (S. Banerjee 2012) celebrated the values of "martial prowess, muscular strength and readiness to go to battle," alongside moral fortitude. The celibate, masculine Hindu "soldier" and the "warrior monk" continue to energize the key cultural expression of nation and manliness within Hindu nationalism (McDonald 1999).

While the British relegated homosexuals and oriental Indians to the position of sexual "others," Indians created their own sexual others. In chronicling British colonial rule in India and the emergence of Hindu nationalism, Prathama Banerjee (2006) argues that Indian elites reacted to the British by producing themselves (elite Indians) as the normative, modern ideal Indians while characterizing Indian tribal groups as the primitive and sexual other. In examining colonial Bengal, she shows how through two processes, one historical and one anthropological, elite Bengalis created a new archetype portraying the Santal tribe as "an allegedly hardworking, sensuous, body centric primitive"—and then rooted them in the past, as less evolved. Thus, elite Indians created their own "primitives" as they constituted their own modernity. In particular, in reacting to British categories of Indians as their sexual others, Indians focused particularly on gender, sexuality, and family as important elements of social reform (Sangari and Vaid 1989; Butalia and Sarkar 1995; Mani 1998; T. Sarkar 2001; Devare 2013; Kapur 2005). Again, there is no easy subjugation story here. The social reformers did not all relegate women to specific roles in the household. In the Arya Samaj movement, for example, women were urged to become good *pracharikas*, or teachers, who had a duty to spread literacy to other women (Allender 2016). Similarly, the Singh Sabha movement among Sikhs and some initiatives within Islam promoted women's education. This was bionationalism, yes, but also something more.[4]

Further, embedded in the reformers' discourses were revised precolonial histories with fresh interpretations of the ancient scriptures, the Upanishads and the Vedas (Bose and Jalal 2011). Hindu nationalism countered Western superiority and white supremacy with a "grand Hindu past." Both *Hindus* (a term for people who lived east of the Indus River) and modern Hinduism must be understood as a particular political project emerging from the complex histories of South Asia (A. Roy 2016).

In a parallel argument, Paola Bacchetta (1999) traces Hindu nationalism's own primitives and sexual "others." While the British rendered Indians as primitive, effeminate, and sexually depraved, Hindu nationalists posited the same characteristics for India's non-Hindu others. As a result, all non-Hindus in India were rendered queer (encompassing nonnormative gender and sexualities). By this move, queer Indians automatically "originate" outside India. Thus, Hindu nationalists posit a "xenophobic queerphobia" by claiming that queerness is not Indian and that the British and Muslims brought homosexuality to India (Bacchetta 1999: 15). These racialized projections of queer excess continue to resonate widely in majoritarian Hindu India (Puri 2016).

Finally, in rousing its Hindu soldiers, key RSS leaders Vinayak Savarkar, Keshav Hedgewar, and Madhav Golwalkar carefully constructed their writing and speeches to highlight and construct an ideology that emphasized the ferocity of Muslim men and women, the weakness of Hindu males, and an ever-present Muslim threat to Hindus and Hinduism. Taken together, these provided a powerful call for Hindu revival, a potent incitement for retaliation against Muslims, and preemptive sexual violence against Muslim women (M. Kumar 2016). This script remains hauntingly familiar.

ACT 4. RELIGION, SCIENCE, AND THE DISCOURSE OF "NATURAL SEX"

We have so far explored the braided genealogies of the sexual. Starting with IPC 377, which until recently criminalized "unnatural" sex in contemporary India, we traced how Christianity, the biological sciences, and colonial expansion brought Victorian sexual morality as normative biology. We saw how the sexual of British colonialism ended in tense and uneven formations in India, even as Hindu nationalists created their own sexual others through positing all non-Hindus as homosexual others. Here, I examine how these braided histories shape contemporary discourses of unnatural sex in India, in particular those that were encoded in IPC 377. Within Hindu nationalism as well as Indian society, the ideal of "marital, procreative, domestic" sexual activity is the only sex legitimized. All other sex is considered illegitimate (Bose and Bhattacharya 2007; Purkayastha 2014). As Puri (2016) argues, understanding Section 377 as a statute that extends beyond homosexuality allows us to understand sexuality in India beyond the homosexual/heterosexual binary that is often presented. Kapur (2007: 235) writes,

> Given the sexophobic content that structures our reality, alternative sexual desire . . . is understood as anything from public kissing between a man and a woman to the presentation of lesbian lovemaking on screen. It is [this] broader phobia regarding sex that sets the framework within which the controversies around alternate sexuality can be understood.[5]

As we saw in the genealogies of the sexual in India, nationalism is key to understanding heterosexism. Homosexuality is not only the sexual "other" but also antinational. Same-sex love means complicity with the West (Bandoyopadhyay 2007). Such essentialized national imaginaries that compelled anticolonial struggles continue to inform the biopolitical regulations of the postcolonial state, one that remains predominantly Hindu in spirit (Bagchi 1990). IPC 377 sanctified ideals of Indian sexual norms through a very narrow definition—man-woman penile-vaginal intercourse. Like in biology textbooks, the "order of nature" is procreative sex, and erotic pleasure without procreation is labeled "unnatural" and fit for repression (Purkayastha 2014). Even after the historic Supreme Court ruling, many Hindu nationalist leaders did not comment publicly or continued to pathologize non-normative sexualities as indications of "mentally sick people" or as "genetic disorders" (Dutt 2018; Gettleman, Schultz and Raj 2018). Two examples powerfully illustrate how the scientific and religious cohere.

The Eugenic Histories of Hindu Nationalism

Scientific theories of eugenics were deeply intertwined with ideologies of Hindu nationalism. Eugenics was without a doubt a broad-ranging project. It was less about a clear set of scientific principles and more about a "modern" way to discuss social problems in biological and scientific terms, spanning the political left and right (Dikötter 1998; Bashford and Levine 2010). Here, I am focused more particularly on eugenic scripts that fed nativist anxieties. The "unnatural" became a placeholder for those deemed undesirable. Indian nationalists embraced eugenic logics into their calls for social reforms. Fischer-Tiné (2006) traces how "modern" Victorian and Edwardian scientific discourses were appropriated and assimilated into overtly anticolonial and nationalist struggles. The scientific discourses of Darwinism, and the subsequent rise of social Darwinism, began in the mid-nineteenth century. These theories fed growing eugenic scripts (Subramaniam 2014) propelled by hysteria about whites' low fertility rates and a fear of white "race suicide." As a result, eugenic programs were largely structured to curtail the reproductive rights of those deemed unfit. With calls for "efficient" use of national resources, eugenic programs and policies were instituted. Eugenic logics focused on producing a superior society by regulating reproductive landscapes—promoting the reproduction of some groups (positive eugenics) while curtailing the reproduction of others (negative eugenics). These scripts were fundamentally about race and class in the West. In India, race found a dubious linkage in caste (a link that never severed), alongside class and religion. Groups were dubbed to have certain proclivities and characteristics. In India, for example, Bengalis and South Indians were believed to be servile races, lacking the vigorous qualities needed for military service. These ethnic groups, sometimes referred to as races, were deemed effeminate. Non-Hindus were deemed sexual others. In particular, the British targeted child marriage, and its promotion of sexual activity in early youth, as a particular ill: "As regards race, there can be little doubt, that the marriage of children, often with aged males tends to the physical deterioration of the human stock, and physical deterioration implies effeminacy; mental imperfection and moral debility" (British doctor in the Indian Medical Service, quoted in Fischer-Tiné 2006).

Hindu social reformers willingly accepted these diagnoses. The data from the decennial census fed the fears of the deterioration, degeneration, and eventual elimination of the "Hindu race" (when other groups did not fare as badly). Hindu nationalists had their own fears of "race suicide." The birth of frail and weak offspring presented "the sad spectacle of prematurely worn out and early dying Indian manhood" (Fischer-Tiné 2006: 249). Swami Dayananda, a key figure, was convinced that Hindus had reached the final stage of a gradual decline (Fischer-Tiné 2006). This crisis energized a renewed focus on bodily discipline instituted by Hindu nationalists. However, as we saw earlier, it is a bifurcated logic: a "muscular masculinity" (S. Banerjee 2006) alongside a promotion of "brahmacharya," or celibacy, by pathologizing Western perverted hypermasculinity. As one writer (quoted in Fischer-Tiné 2006: 251) argues, British education

> takes no notice of the body, or which does not enable a person to control his passions and do good deeds must be extremely injurious. . . . An English educated person is generally victim to bodily diseases: headache, indigestion, and nervous debility being the most common complaints with him. Its other effects are, that students learn expensive ways of living and attach too much importance to ceremonial forms.

This scientific genealogy of eugenics shaped Hindu nationalism during British colonial rule. These logics continue in family planning policies in colonial India, when racial, ethnic, and caste differences grounded eugenic policies (Hodges 2008).

Discourses in the Courts

To understand the intersections of science, religion, and nationalism in IPC 377 and constructions of the unnatural, the history of recent legal cases in the Delhi High Court in 2009 and the Supreme Court in 2013 and 2018 are worth examining. The many genealogies we have traced help shape how ideas of the natural merge with the moral and ideas of science with religion. Five key issues help us understand how biopolitical arguments before the Supreme Court asserted homosexuality as "unnatural" and "emasculating."

First, in the legal discourses that surrounded the case, "carnal intercourse against the order of nature" was defined and interpreted as any penile penetration/insertion with the intention of satisfying "unnatural lust." Acts included in this interpretation extend beyond sodomy to insertion between thighs, holding tight in hands to satisfy (unnatural) bodily appetites,[6] and also a case of extreme cruelty of a husband toward his wife (Mandal 2016).[7] The conception of "carnal intercourse" thus legislates a vast repertoire of sexual interactions by reducing "natural" sex to penile penetration/insertion for procreative sex within the bounds of heterosexual marriage. Leading advocates pronounced any act that was non-procreative, pleasurable, or an "imitation" of heterosexual procreative sex as "sexual perversity." These acts were deemed analogous to other "unnatural" categories, such as bestiality and sex with minors.[8] Having traced the roots of the unnatural and natural in early Indian nationalism and biological discourse, we should not be surprised by this definition. The descriptions of "unnatural sex" and "unnatural lust" reinvoke the scientific imaginations of natural and normative as well as the puritanical sexual politics that shaped early nationalism in creating "non-Hindus" as subjects who were simultaneously also "sexual others." This move simultaneously constituted "normal" citizens whose natural lust and "normal" sex were the paragons of a Hindu India, upholding traditional heterosexual, patriarchal norms. The success of the LGBT rights movement in India (and other countries) has been grounded in "unqueering" gay sex i.e., normalizing "gay sex" as being as normal and natural as "straight sex," and particularly setting it apart from other "unnatural sex" that should continue to be seen as deviant, perverse and "unnatural." Indeed the Supreme Court's ruling very specifically only decriminalized consensual, private "gay sex"; it left IPC 377 in place as a criminal statute for other kinds of "unnatural" sex such as sex with children, bestiality, and other "unnatural" acts. The legacy of the "unnatural" lives on.

Second, the broadened definition of the unnatural relied on anatomical "correctness" of the body and assumptions about how organs were "naturally" supposed to be used. Again, the history of biology helps us understand the biological logic of assigning a purpose to all organs, and especially how the reproductive system came to be marked as the critical system for biology. The primacy of the reproductive system and its attendant anatomy and functionality re-emerges in these debates. For instance, S. Radhakrishnan, a lawyer in favor of reinstating Section 377, invoked *Black's Law Dictionary* to distinguish *natural* from what is artificial and contrived, stating that the "basic feature of nature involved organs, each of which had an appropriate place. Every organ in the human body has a designated function assigned by nature. The organs work in tandem and are not expected to be abused. If it is abused, it goes against nature."[9] Additional solicitor general P. P. Malhotra, representing the Union Home Ministry, echoed this argument as he conflated the unnatural with the immoral. In response to Justice Singhvi's question on "what is immoral," Mr. Malhotra argued, "Nature has made man and woman. His penis can be inserted into female organ because it is constructed for that. It is natural. Now if it is put in the back of a man where human waste goes out, the chances of spreading disease is high. There are UN studies to show this."[10]

Third, the arguments show that those acts deemed unnatural, and the bodily logic of why they are deemed such, are explicitly tied to the need to maintain heterosexual, heteronormative order. Most advocates arguing for the reinstatement of Section 377 cited the dangers for "Indian social structure and the institution of marriage." Lawyers also pointedly quoted from Hindu mythology as evidence for 377 to be upheld on religious grounds. One lawyer quotes from the *Mahabharata* to implicitly argue that revoking Section 377 would create "bad conduct" and more "desire" and would destroy notions of motherhood and fatherhood, and hence the greatness of India:

Four questions were posed to Yudhishtira by Yaksha,
and the answers are:

Q: What is more nobly sustaining than the earth?

A: The mother who brings up the children she has borne is nobler and more sustaining than the earth.

Q: What is higher than the sky?

A: The father.

Q: What is Happiness?

A: Happiness is the result of good conduct[.]

Q: What is that—by giving up which, a man becomes rich?

A: Desire—getting rid of it, man becomes wealthy[.]

He continued, "This cherished concept of Father, Mother, Happiness and Wealth, will receive a lethal blow from homosexuality and gay marriage. Thus, this ancient land will lose its nobility and rightful and honoured place in the world. . . . These cherished concepts go back 5000 years."[11]

These assertions of homosexuality as unnatural and destructive of the primacy of heteronormative social structures also highlight that it is emasculating. Lawyers argued that "Carnal intercourse against the order of nature" had to be criminalized. If not, "such acts have the tendency to lead to unmanliness and lead to persons not being useful in society."[12]

HIV/AIDS also played a prominent role in adjudicating whether 377 should be revoked or reinstated. They framed queer populations either as victims vulnerable to risks of sexually transmitted diseases and sexual assault or as villains threaten public health. While both sides cited 2005 data that found greater HIV/AIDs prevalence in the MSM (men who have sex with men) populations (8 percent) compared to the general population, arguments in favor of decriminalizing homosexuality presented detailed accounts of how Section 377 hindered HIV/AIDS prevention efforts.

Echoing the points on the anatomically "correct" use of organs to distinguish natural versus unnatural acts, a lawyer representing the Apostolic Churches Alliance (ACA) and Utkal Christian Council questioned whether the Delhi High Court judgment would help with HIV/AIDS prevention, citing a Centers for Diseases Control (CDC) study to claim anal sex as riskier: "The anus is vulnerable due to tears due to anal sex which influences the likelihood of getting AIDs." He also mentioned "a number of health problems resulting from anal sex, like diarrhea and gay bowel syndrome."[13] Even as this lawyer included the caveat that homosexual acts, not sexual orientation, lead to public health problems, the large body of arguments treated homosexuality as a public health problem and a disease. The lawyer representing the Union of Home Ministry argued that "because of their risky sexual behavior, MSM and female sex workers are at a high risk of getting HIV/AIDS as compared to *normal human beings*" (emphasis added). Another lawyer went one step further, deeming homosexual sex as unlawful due to its negligent spread of disease: "It is very difficult to identify gay people, homosexuals, sex workers. Like for Malaria, Cholera . . . we need a rehabilitation program. This will spread the disease otherwise. For the prevention of smoking and addictions there already are clinics. NACO [the National Aids Control Organization] as well as [the] Naz [Foundation] are misdirecting themselves."

The arguments in the Supreme Court show a persistent attempt to establish heterosexism not only as normative sexual behavior but as part of national culture. The "authentic Indian" could not be homosexual. The assistant solicitor general concludes: "*In our country, homosexuality is abhorrent* and can be criminalized. . . . Article 19(2) expressly permits imposition of restriction in the interest of decency and morality. Social and sexual mores in foreign countries cannot justify de-criminalisation of homosexuality in India . . . *in Western societies, the morality standards are not as high as in India*" (*Naz Foundation v. Government of NCT of New Delhi* 2009: 23, emphasis added). Another counsel is quoted as saying:

> Mr. Sharma submitted excerpts from the Manusmriti, the Bible and the Quran to the Bench. . . . Mr. Sharma also referred to Mahatma Gandhi's disapproval for "unnatural vices" in 1929. . . . Mr. Sharma further submitted that homosexual sex was unnatural and immoral and the Indian society abhorred such perverted practices. . . . Mr. Sharma contended that there was no concept of sexual minorities in the Constitution of India. . . . The Law had clearly laid down what was natural and not natural in accordance with the prudence of an ordinary man. (*Koushal and Ors. v. Naz Foundation (India) Trust and Ors* 2012: 22–23)

The Supreme Court itself reinforced this line of thinking when it first ruled in 2013: "We have grave doubts about the expediency of transplanting Western experience in our country. Social conditions are different and so also the general intellectual level. . . . In arriving at any conclusion on the subject, the need for protecting society in general and individual human beings must be borne in mind" (quoted in Shandilya 2016). The debates around 377 feel anachronistic at best, evoking language, rhetoric, terminology, and scientific and religious discourses from another time. In an act of irony, the Supreme Court first ruled on Section 377 close in time to a second ruling on the legal recognition of transgender people. This second case petitioned the "inclusion of a third category in recording one's sex/gender in identity documents like election card, passport, driving license and ration card; and for admission in educational institutions, hospitals, access to toilets, amongst others" (Lawyers Collective 2013). Here, the Supreme Court ruled in favor of a third category of sex/gender. In reading the two decisions together, Svati Shah (2015) astutely points out how "in a strange twist, transgender people in India, and especially hijras, now have the right to official recognition as members of a third gender, but do not have the right to have 'unnatural sex.'" At one level, Shah's observation emphasizes the point that what is being policed primarily here is "unnatural" sex and not disruptions of the gender binary. At another level, the two rulings underscore how a third category of sex refuses to recognize and accord transpeople the right to the gender they wish to embody. Here, binary gender is reinscribed.

ACT 5. RESCRIPTING THE SEXUAL: CULTIVATING POSTCOLONIAL IMAGINARIES

In feminist theorizations, sexual pleasure is almost exclusively associated with the West (Shandilya 2016). In contrast, other than the pathologized sexualities of the porno-tropics, the third-world oriental subject is "constructed almost exclusively through the lens of violence, victimization, impoverishment and cultural barbarism" (Kapur 2009: 382). The mythological stories of India present a lush sexual landscape full of possibilities. These stories destabilize any normative notions of sex, gender, or sexuality. In these imaginations, everything is natural and possible. Yet despite this narrative backdrop, we have seen how Hindu nationalism evolved a narrow procreative, heteronormative, and domestic model of the sexual. The genealogies of the sexual in India force us to reckon with vastly contradictory threads of history—one where sexuality is fluid and multiple, and another where it is rigid and binary.

While transcripts and debates surrounding IPC 377 feel anachronistic, there is also ample evidence that more liberatory imaginations had weathered alongside. It is clear that Hindu nationalists do not speak in one voice or represent all Hindus. Some Hindu religious leaders argued for a more capacious and permissive reading of history and Hinduism. Simultaneously, debates within evolutionary biology about the primacy of sex and sexual selection have also grown. These two recent strains open up new religious and scientific imaginations.

While the BJP's official stance supported IPC 377, several Hindu nationalists in the RSS and BJP spoke in favor of decriminalization for a variety of reasons. One argued, "I don't think homosexuality should be considered a criminal offence as long as it does not affect the lives of others in society" (*Hindustan Times* 2016). Others pathologized homosexuality by labeling it a disease rather than a crime—it was a "psychological disease" or "a genetic disaster" (Anasuya 2016) that might even be "cured" (D. Nelson 2009). Although some BJP and RSS members initially supported decriminalization, soon the party position in 2013 rallied around the more traditional voices.[14] Some religious leaders who initially supported decriminalization fell in line with the main party. One prominent Hindu nationalist religious leader, Baba Ramdev, has made claims that homosexuality is an addiction that is "aprakritik" (unnatural). In keeping with the newly emerging links between Hindu nationalism and capitalism, the entrepreneurial Baba Ramdev founded Patanjali Ayurved, and he claims that some of the business empire's many herbal products (derived from the wisdom of ancient Hindu medicine) will "cure" homosexuality (Madhukalya 2013).

However, other religious leaders strongly supported decriminalization. For instance, Sri Sri Ravi Shankar of the Art of Living Foundation argued for an interpretation of divinity, consciousness, and love that supersedes genetics and allows for fluidity in gender and sexuality. In response to a question on whether homosexuality is a sin, he said,

> See, these tendencies come in you because you are made up of both mother and father, a combination of two. Everybody has both male genes and female genes. When male chromosomes are dominating or when female chromosomes are dominating, these tendencies come and go. You don't have to brand yourself about that. It can all change. There is a possibility of those tendencies changing. Know that you are more than just the body. You are a scintillating consciousness, energy. You are sparkling light, and so, identify yourself with that light more.[15]

Ravi Shankar publicly condemned Section 377, maintaining that "Hinduism has never considered homosexuality a crime." In 2018, the Hindu nationalist government chose to deliberately not take a position, and even after the ruling, leaders were largely silent. The Modi government "sat precariously on the fence, trying to appease both its hardline conservative base, as well as the burgeoning youth vote" (Dutt 2018). Finally, it is worth noting the increasing role of the liberal idea of "individual" freedom in recent Supreme Court rulings – from individual rights to love, and individual rights to privacy.

Strikingly, the position pathologizing homosexuality is one that is common to both the Hindu Right and Muslim and Christian evangelicals. The lawyer representing the side in support of reinstating Section 377, Hufeza Ahmadi of All India Muslim Personal Law, declared there were no religious grounds to support homosexuality. He stated that homosexuality is condemned by the Bible, the *Arthashastra*, the *Manusmriti*, and the Quran. Much like the Hindu Right, which seeks to reform Hinduism away from Western narratives, Ahmadi is dismissive of the prevalence of homosexuality in precolonial India: "the fact that homosexuality was accepted in precolonial India, the source is the diary of an English officer. Western writers puff up the mysticism of the east."[16] And yet the idea of sexual offenses "against the order of nature" per se is specifically a British one that was transplanted to India in colonial times.[17]

Such colonial legacies translate into the management of postcolonial bodies, reinscribing colonial imaginations into our view of postcolonial biologies. Yet in this history, we also see that colonial legacies do not fully map onto Indian conceptions of the sexual. Against the backdrop of a vibrant and lively queer genealogy of a fluid and open sexual politics in Indian history, we see in contemporary Hindu nationalism a reinscribing of the complex genealogies of the sexual. Similarly, within biological thought the primacy of neo-Darwinian theory, which posits individual organisms, sex, and sexuality as the sole generators of variation and evolution, is also gradually eroding (Margulis and Sagan 2003; Margulis 2008; Roughgarden 2004, 2005, 2009). The history of the sexual in India and the vibrant stories of Indian mythology also hint at sexual animacies (M. Chen 2012) beyond the human. Taken together, these modes displace the privileging of sex and sexuality as the key engines of evolution, to usher in queer biologies beyond sex, opening up the tremendous possibilities beyond the binaries of natural and unnatural, and the logics of male and female and the heterosexual couple.

◆ AVATARS #3 ◆

The Story of Nādu and Piravi

For decades now, Nādu has watched the growing nationalism on Kari with deep concern. The vantage point was all about *me, me, me!* For some, it extended into the possessive: a focus on *my* son, daughter, wife, husband, partner, parent, friend, family, community, nation—and all that they owned. Nādu was the node, or Kanu, of place and space. An amorphous ball of magenta, Nādu was capable of becoming as tiny or gigantic as it wanted. It could spread out across enormous spaces or shrink into a small crevice. Nādu enjoyed the thrills of growing and shrinking, of inhabiting forms that were enormous and forms that were miniature, both deeply pleasurable in their own ways. Nādu rolled into Uruvam's cuboidal black body and coursed through to sample the new results from Kari. *Disappointing*, it thought. Despite Nādu's best efforts in coordination with Uruvam and Amudha, *Homo sapiens* seem to perennially want more place and space, and nationalisms and conquering armies were forever usurping lands and fighting over them. Through small avatars, Nādu had introduced new technologies and ideas of the commons, of sharing economies, of community and social movements. Some of these were discovered, but eventually the forces of greed seemed to take over. In biological thought, Nādu introduced ideas such as cooperation, mutualism, symbiosis, synergy, symbiogenesis, sympoiesis—all signaling the necessity of collective action for planetary survival. Nādu was always telling the liquescent Amudha that at least humans could not claim that sharing and collaborating were alien traits! There were plenty of examples.

Things had been going well on Kari until *Homo sapiens* appeared. Until then, Nādu's ideas had reverberated through the planet. Plants and animals took various forms. Some were tiny, others large, and for a while they roamed the planet freely. Nādu and Uruvam loved to scan the varied shapes and sizes on the planet. But humans appeared to have developed a deep belief in acquisition—collecting and acquiring people, objects, and things; they wanted to possess anything and everything. Their greed appeared insatiable. Nādu had grown concerned. How do you change that? Over the decades, it had started introducing new material changes. The planet grew warmer. Ozone levels rose, and ultraviolet rays ushered in increasing levels of mutations. With time there were global changes in Kari's climate. Hurricanes, typhoons, earthquakes, tsunamis, cyclones, floods, and droughts became more common. Agriculture belts began to shift. It was clear that all earthlings shared a common atmosphere; it flowed freely across nations. Amudha worked on chemicals flowing through the rivers—all nations needed water, after all. *Might their efforts bring about greater cooperation?*, Nādu wondered. No nation could solve the problem alone, since the environment was common and shared by all.

What would humans do?, Nādu wondered. Throughout life on Kari, creatures had embraced Nādu's impulses and moved freely where they wanted to and could. Humans had interrupted this flow by creating nations and guarding their borders. *What audacity*, Nādu had thought. What hubris to appear on a new land and just declare it your own! It was especially outraged that some would steal someone else's land, claim it as their own, and then proceed to eliminate the original inhabitants. It was outrageous for humans to believe that they had permission to commit genocide, Nādu fumed. The settler colonists across the planet were always Nādu's deepest lament of all the creatures it had imagined on Kari. The horror and havoc they had unleashed! Predictably, after declaring new nation-states, the powerful humans in those nations began building walls and refused to let any creature in. Most frustratingly, they began considering themselves native to the land and began pogroms to eliminate all creatures—plant, animal, and human—that were "foreign." So afraid were they about contamination that they created "seed banks" and immortalized "pure" seeds for posterity. Nādu and others witnessed with horror the widespread exterminations, the lost souls filling up Avatara Lokam. But then of course,

Nādu had the last laugh. With fast-changing environments all over the planet, no creature was well adapted to its environment any longer! So Nādu sent lost species back to the planet, into new and interesting locations, and they thrived. Humans could not pull them out quickly enough, and settler colonists grew discontent.

Green purity brigades called *Earth lovers* campaigned for clean land, air, and water. For a while it seemed the world was listening, but suddenly all was undone. Emperors of the White Brigade of the West, and the Emperor of the Saffron brigade in the East, and the global Green Brigades blustered and thundered:

Purify the Earth!
Clean the Earth!
More, more, more!
Planet is here for us.
Banish the shrinking violets, and the complaining "snowflakes."

They pushed for purity—of race, blood, religion, nation, nature, species. Who decides? "We will! We are the final arbiters," the emperors bellowed. Their prophecies were fulfilled—there were no criteria or rules, just imperial decree. Then the massive purity campaigns began—throats slashed, guns unloaded, bombs dropped, earth mined, animals slaughtered, plants uprooted. Genocide unleashed across the planet as purity campaigns were waged through royal edicts. Countless innocent victims across the globe cried out to the heavens, "Help us! Save us from the purity brigades!"

Avatara Lokam watched in horror. We can no longer be silent, they decided. Uruvam, Amudha, and Nādu conferred and agreed that it was hopeless. They would need to wake up Piravi, the Kanu of genesis and re-creation.

The long-slumbering Piravi shook itself awake. Its giant purple beak preened its luxuriant red feathers. Its brown tongue smoothed the scales of its orange tail fin. Below its feathers, small white tendrils grew, and slowly they merged with tendrils on the ground. Piravi was ready to trigger a new cycle of life on the planet. As Piravi rose, small Earth beings across the lands and oceans extended their tendrils to connect with one another. Piravi, endowed with the power of *mahima* or limitless growth, began to burgeon. As the tendrils across the globe joined together in unison, Piravi joined them, and soon they enveloped the globe. All across the planet people felt the ground beneath their feet shifting, the oceans rising. Was it an earthquake? A tsunami? They had never seen this before. Piravi rose into the skies, appearing as a giant colorful bird, its flowing red feathers streaking the blue skies, its ferocious eyes glistening and thunderous beak shrieking. The giant bird shook and rumbled, haunted by the disrespect it had endured for centuries. The military launched all its firepower, but Piravi merely swallowed them. Piravi's scaly tail fin flowed into the water and enlarged into the giant seas. Like a giant fish, it enveloped the oceans, rivers, and streams. Naval fleets were sent out, but it swatted them aside. Piravi was formidable, and its multimorphous form took over the planet from air, water, and ground, effortlessly chomping away the military's finest like delicious candy.

Piravi churned the oceans, whirled the skies, and swept the lands. Consuming, digesting, blowing the world about. In Avatara Lokam, Uruvam, Amudha, Arul, Néram, and Nādu readied. Control-Alt-Delete: individuals and species disappeared. Control-Alt-Create: other individuals and species from Avatara Lokam emerged with a new lease on life. Command-Control, and earthly memories would be soon erased.

Earthlings saw a pattern. Those isolated, insular individuals, ideas, and species that had little interaction with the world and lay petrified in their old ways perished. The avaricious, the hateful, the joyless, the selfish, the acquisitive, the individualists, and the unimaginative were particularly targeted. Those curious, open to change, who loved and respected what was around them, who had moved with the times, adapting to new places and new peoples, the joyful, the creative and the imaginative thrived and lived on. The capricious emperors and their false slogans of purity were destroyed.

New species emerged and were welcomed. Some had multiple limbs to navigate new terrains, multiple genitalia to enable new, imaginative offspring, multiple sensory organs that expanded the senses, overflowing genomes filled with promises for the future. These were not deformities, abnormalities, or perversions. They were variations, mutations, adaptations, offering the possibilities of endless innovations. The wise always embraced change.

Yet again, as it had during previous mass extinctions, Piravi reshuffled the world. Kari, reseeded with new possibilities, emerged as a thoroughly impure mutt of a planet, a thriving cosmopolis! Avatar Lokam settled back and erased earthly memories of the historic events.

What did the future hold? With the world reshuffled, might the old ways reappear? Earthly biologies were too malleable, good and evil too contextual, love and hate too entangled. Was carbon-based biology doomed to repeat its genocidal ways? Would humans learn to accept change, welcome the new, adapt to changing climates, and embrace change to imagine new futures? Iteration #1729 was born.

CHAPTER THREE

Return of the Native

Nation, Nature, and Postcolonial Environmentalism

If there is anything that radically distinguishes the imagination of anti-imperialism, it is the primacy of the geographical in it. Imperialism after all is an act of geographical violence through which virtually every space in the world is explored, charted, and finally brought under control. For the native, the history of colonial servitude is inaugurated by the loss of locality to the outsider; its geographical identity must thereafter be searched for and somehow restored. . . . Because of the presence of the colonizing outsider, the land is recoverable at first only through imagination.

EDWARD SAID, *Culture and Imperialism*

Postcolonial cultures' reliance on myth and local legend is an effort at de-contamination, a process of freeing their cultures from colonialism's pervasive influence. The return to roots—while running the very real danger of fundamentalism, reactionary nativism, and chauvinism—is an attempt to gain a measure of self-affirmation that is not tainted by colonialism.

PRAMOD NAYAR, *Postcolonial Literature*

LIVING IN A WORLD OF STORIES AS A YOUNG GIRL, I REMEMBER being struck by the many narratives of rescue that surrounded me—nationalist narratives of India's rescue from colonial rule, religious stories of Rama rescuing Sita in the *Ramayana*, the intrepid heroes of Bollywood

113

rescuing their damsels in distress, and the global exploits of environmentalists rescuing the splendid biodiversity of our planet. The objects of the rescue were always abject and always rendered feminine. The *Ramayana* in particular provided much fuel to my budding feminist sensibilities.

In Hindu mythology, Rama is the seventh avatar of Vishnu and the idealized hero of the *Ramayana*, a central epic in the Indian imagination. While there are multiple versions and reinterpretations of the *Ramayana*, the most common version, and the one I grew up with, was about King Dasaratha of Ayodhya and his three wives. When the king chooses Rama to be his heir, Rama's jealous stepmother Kaikeyi, reclaiming the two boons the king had granted her, demands that Rama be exiled into the wilderness for fourteen years and that her son, Bharata, be made king. The king and Rama, being righteous rulers, keep their promises. Sita, Rama's wife and the paragon of wifely virtues, gives up the luxuries of royalty and follows him.

While they are in the forest, the demon Ravana abducts Sita (a revenge plot that begins with Rama and his brother Lakshmana spurning Ravana's sister). In an attempt to rescue Sita, Rama needs to cross the strait between India and Sri Lanka. He entreats the help of the monkey god Hanuman, who, along with his monkey army, builds a stone bridge that allows Rama, Hanuman, and both their armies to cross the seas to rescue Sita. During the war, Lakshmana is wounded, and Hanuman (who can fly) is sent to fetch four herbs: *mruthasanjeevani* (restorer of life), *vishalyakarani* (remover of arrows), *sandhanakarani* (restorer of the skin), and *sarvanyakarani* (restorer of color) (Balasubramanian 2009). Not able to distinguish the four unique herbs in the multitude of the forest, he flies back with the entire hill! Lakshmana is restored to health. The war is won and Sita is rescued.

After fourteen years of exile, Rama returns to the throne that his brother Bharata has saved for his return, ushering in the *Rama Rajya*, a period of righteous rule by Rama. Subsequently, Rama, on hearing a washerman question Sita's chastity after the abduction, asks Sita to walk through fire to prove her purity. She successfully does so.

The story revolves around the evil machinations of several women, leading to the rescue of others. Growing up in postcolonial India, as a budding feminist I scoffed at these stories. I found the plot distinctly misogynist and the various nonhuman and mystical characters claptrap and nonrational. Revisiting the story now, I see other elements. I am fascinated by this story and its narrative imaginations. As Anil Menon (2012: 3) argues:

> The ancient South-Asians had chutzpah. They imagined our universe as existing for a duration of 311 trillion years (100 Brahma years), about 23,000 times larger than the scientific estimate for the current age of the Big Bang universe (~13.5 billion years). They imagined multiple universes, frothing in the event-sea of creation and destruction. They imagined space and time as being illusory in the absolute and relative across the sea of universes. They imagined consciousness in all of matter, not just human beings. Divinity didn't frighten them. The Rig-Veda expresses doubt on the omniscience of the creator. The ancients imagined weapons that could flatten mountains, unravel minds, devastate entire armies, destroy worlds and even annihilate the gods themselves.

I have also discovered that there are many versions of the *Ramayana*. A. K. Ramanujan's (1991) wonderful essay on the three hundred *Ramayanas* chronicles that in addition to the better-known Tamil *Kamba Ramayana* and the Tulsidas *Ramayana*, there are many others, such as the Jaina, Kashmiri, Urdu, and Persian *Ramayanas* (A. Menon 2012). There are also alternate versions: *Sitayanas*, or feminist versions (Kishwar 1997; Rangarajan 2009; Gokhale and Lal 2010; A. Menon 2012); antiracist versions in the South, where the dark-skinned demon king is renarrated as an erudite and righteous character (Jaffrelot 2008; Viswanath 2017); and queer readings of sexuality in the epic (Mangharam 2009). Recent biological work on plants and the recovery of nonhuman worlds in feminist and indigenous thinking highlight the *Ramayana*'s imaginative work. I am moved by the rich tapestry of nonhuman agency that imbues forests and their many inhabitants with narrative powers. Sita is the daughter of Earth (Bhumi Devi). In one version of the *Ramayana*, Sita is so frustrated with repeated doubts about her fidelity that she appeals to her mother and is swallowed back into the ground. The epic effortlessly mingles naturalistic and animist sensibilities in its narratives. Sovereign power is imagined over all living beings and landscapes. In the *Rama Rajya*, glorious rains arrive on time, trees thrive, and the natural order flourishes under a *dharmic* ecology (Gopalakrishnan 2008; Rangarajan 2009).

With Hindu nationalism's rise, the power of Rama has also risen. If there is one god Hindu nationalism coheres around, it is Rama. Valmiki, author of one version of the story, the *Valmiki Ramayana*, remarked, "As long as mountains and streams endure upon earth, so shall the Rama's story continue to circulate in the world" (Paddayya 2012). This was prophecy indeed. Rama lives on. We have seen the violent politics of the Ram Janmabhoomi movement in Ayodhya, the city Hindu nationalists insist is Rama's birthplace. Here, we witnessed the destruction of a mosque, the Babri Masjid, in 1992 and Hindu nationalists' insistence that a Rama temple be built in its place (Noorani 1989). A legal case is ongoing in the Supreme Court.

We have also witnessed controversy around Rama Setu, Rama's Bridge (also known as Adam's Bridge), the focus of this chapter. This political drama stages moral injustice and a politics of rescue. The violent takeover of land and its subsequent reclamation remain important elements of the *Ramayana* and of India. Geography and space, as Edward Said (1993) reminds us in the opening epigraph, are important sites in the colonial imaginary. In this context, Rama emerges as a powerful, contemporary figure; election campaigns have drawn direct parallels between Lord Rama and Prime Minster Modi (Ghosal, Sheriff, and Kaushika 2014; *Economic Times* 2014). In the promise of *Rama Rajya* is a promise of a renewed Hindu world: its governance, landscape, ecology, and people.

The gendering of space is a haunting continuity from colonial to postcolonial approaches to space and geographies. The nation and nationalisms remain resolutely gendered (McClintock 1993; S. Banerjee 2003; de Mel 2002). Images of Mother India, or *Bharat Mata*, remain strong symbols of the nation (Ramaswamy 2010). As feminists have long recognized, the nation, eternally female, is rendered passive and infantilized in this rhetoric, invoking protection and care even while violence is spilled in her name. Nature, like nation, is also resolutely female. As we talk of the nation as mother, we talk also of "mother nature." Feminists have argued that patriarchal societies render nature, like the nation, as passive and exploitable (Merchant 1981; Keller 1985). In reclaiming the "other," some feminists have extolled the feminine as the only path to a sustainable planet, arguing that the "feminine principle" is linked with diversity, and a return to feminine principles of community and sharing offers the only hope of a sustainable future with a creative and organic nature (Shiva 1988). In this sense, all postcolonial environmentalism is a form of gender politics. The promise of Rama is the promise of rescue and return. The past is prologue.

NATIVISM, ORIENTALISM, AND THE POLITICS OF RETURN

I begin with the story of the *Ramayana* because Rama and Hanuman have emerged as key Hindu gods of a robust and muscular nationalist masculinity in contemporary India but also because Rama's story powerfully underscores the power of nativism and the politics of rescue, and return. The Bajrang Dal, the militant organization associated with the Hindu nationalist Vishva Hindu Parishad (VHP), celebrates Bajrang, or Hanuman, and has emerged as a potent site of a militant Hindu nationalism (Bhownick 2018). In the *Ramayana*, Rama was unjustly sent out of his home into the forests. But as the true native, he is ultimately victorious and returns to his rightful place on the throne to usher in a righteous and utopic future for his native kingdom. Nativist rhetoric is powerfully marked by "right"— the rightful and righteous. It implies a natural order, something preordained, something that was meant to be. The "meant to be" inevitably harkens to an idyllic mythic past that was lost because of injustice or because of the arrival of the foreign, the colonizers who seized land. The injustice of the usurpation of land, nature, nation, and memory is highlighted. The *Ramayana* endures as a powerful contemporary force in part because its narrative highlights the injury of exile and the righteousness of return. These are potent tropes in nationalist politics, helping mobilize a nation through an affective politics of injury and nativist return. Hindu nationalism is not alone in using the politics of nativism (Subramaniam 2014). Many have done so, including some environmentalists. Here we examine two recent cases to explore how nativism, orientalism, and anticolonialism bring together the power of nation and nature through a politics of return. Act 1 explores the controversies around a shipping canal and a bridge, the Rama Setu. This case highlights the power of nativism and an ancient idyllic mythic past in contemporary India. Act 2 explores the power of nativist values that exalt the past as not only idyllic but also wise and just. It shows how science embraces orientalist ideas of the "East" to mythologize India as a site of ecological wisdom and justice. Both acts highlight the power of nativism as a trope of the ecological and environmental sciences—a deep sense of injury that globalization has wrought, and that nativism will solve with the violent expunging of the nonnative from nativist borders. The chapter concludes with an exploration of the power of a politics of injury and righteous return and a broader examination of postcolonial environmentalism in contemporary India.

ACT 1. RESCUING NATION AND NATURE: THE SETUSAMUDRAM SHIPPING CANAL PROJECT

Since independence, India has imagined itself as a modern state, embarking on a Western development agenda in which nature is a resource to be utilized (Prakash 1999: 234). Like in development nationalism, cultural and political aspirations of nature conservation and environmental protection come together in a politics of "ecological nationalism" (Caderlof and Sivaramakrishnan 2006). Ecological nationalism has included the construction of megaprojects including dams, coal and hydroelectric power plants, and more recently nuclear power plants. With growing industrialization, infrastructure development has become important. One such project included a shipping canal on the southern border of India. The Setusamudram Shipping Canal Project (SSCP) involved the creation of a passable channel between India and Sri Lanka for commercial ships, from the Arabian Sea and the Gulf of Mannar in the west to the Bay of Bengal in the east (N. Menon 2013). The Indian government had sought and approved construction of the SSCP, but after construction began, two groups—religious nationalists and environmentalists—vigorously and vociferously opposed the project. The controversy arose from the need to destroy a chain of underwater structures in order to create the channel. Were these structures natural limestone shoals, produced in the last Ice Age, linking Mannar Island in Sri Lanka and the island of Pamban in the Indian subcontinent (Adam's Bridge)? Or were they the Rama Setu, the bridge built by Hanuman and his monkey army as narrated in the *Ramayana* (*Encyclopedia Britannica* n.d.a)?

Feagans (2014) provides a useful history of the SSCP. Since the eighteenth century, many projects have proposed such a canal.[1] The current project began in 2004 after the Hindu nationalist Bharatiya Janata Party (BJP), then in a coalition government with the National Democratic Alliance (NDA), approved a budget to begin construction (Jaffrelot 2008). The first concrete steps, however, began during the subsequent government of the secular, Congress Party–led United Progressive Alliance. The SSCP was finally approved and funded in June 2005 with little debate, and Setusamudram Corporation Ltd. was created (*Economic and Political Weekly* 2005). Then prime minister Manmohan Singh of the Congress Party inaugurated the project on July 2, 2005 (Manmohan Singh 2005). Dredging began in December 2006 (*Hindu* 2006) but was halted in September 2007 over Adam's Bridge and in July 2009 in the Palk Strait (Feagans 2014).

Hindu leaders proclaimed that Rama Setu was the bridge built by Lord Hanuman and his monkey army over a million years ago so Rama could rescue Sita (*Times of India* 2002; O'Connor 2007; N. Menon 2013). While historians have long refuted the *Ramayana* as history (Basham 1954), debates over its veracity erupt at regular intervals as academics explore potential locations and objects from the *Ramayana* in the real world. For example, scholars look at genetic affinities to try to locate tribes mentioned in the epic (Chattopadhyaya 1975; Nagar and Nanda 1986; Konduru 2016; Chaubey et al. 2015). Botanists look for potential extant plant species. Whether to prove the veracity of the epic or to look for potential medicinal herbs that are "worth probing lest we miss out a potentially important biological resource" (Ganeshaiah, Vasudeva, and Shaanker 2009: 484), these efforts demonstrate the epic's continued link with contemporary scientists. One focus has been a search for the plants Hanuman brought to save Rama's brother Lakshmana. Projit Mukharji (2014), in his fascinating article on *vishalyakarani*, one of the named herbs, traces the long history of such explorations as "retro-botanizing." He relates how one of the plants identified as *vishalyakarani* was *Eupatorium ayapana*, which on further examination turned out to be a Brazilian weed that been introduced into South Asia in the modern age. Despite this discovery, the herb's use as a medicinal product continued. New research that began in the mid-1930s at the Bose Institute, a flagship center of nationalist Indian science, eventually relegitimized *vishalyakarani*'s medicinal reputation in the name of "Hindu medicine." Mukharji (2014) reminds us that such explorations are not simple or transparent but instead open up multiple histories, multiple cultural pasts and futures. Yet as he argues, the popular use of this plant produced a robust future for it as a medicinal plant. Botanists have focused on plants that have biochemical properties that resemble the medicinal effect of the plant touted in the *Ramayana* (Antony and Thomas 2011; Sah et al. 2005; Balasubramanian 2009; Ganeshaiah, Vasudeva, and Shaanker 2009). The *Ramayana* thus remains hauntingly relevant, animating attention across a number of sites.

The current controversy over the planned canal was a reprise of a story several years earlier, when Hindu nationalists claimed that NASA photos of the area revealed Rama Setu, the alleged bridge. NASA officials debunked this theory, clarifying that their images could not conclusively prove the origins or the age of the structures and could "certainly not determine whether humans were involved in producing any of the patterns seen" (*Times of India* 2002). Indian scientists and historians contributed to the discussion. Eminent astrophysicist J. V. Narlikar stated that there was no evidence that the bridge in question was the one in the *Ramayana*. Historian R. S. Sharma argued that the *Ramayana* itself was not as old as the shoals, and there certainly was no human habitation in the subcontinent over a million years ago (*Times of India* 2002).

Other players joined the fray. The chief minister of the southern state of Tamil Nadu implored the central government to ignore the "religious fundamentalist forces" (Rediff India 2007). In turn, Hindu leaders mobilized huge numbers of their followers against the desecration of a religious and historical site. In May 2007 the Lok Sabha, the lower house of the Indian Parliament, was paralyzed by the BJP protesting the lack of archaeological evidence. In response, an independent government body, the Archeological Survey of India (ASI), filed an affidavit on September 10, 2007, with the nation's Supreme Court. They concluded that the area in question contained natural formations of shoals and sandbars and not an artificially created bridge (Das 2007). They argued that "The Valmiki Ramayana, the Ramcharitramanas by Tulsidas and other mythological texts, which admittedly form a part of Indian literature, cannot be said to be historical records to incontrovertibly prove the existence of characters or the occurrence of events depicted therein" (Indo-Asian News Service 2007).

This conclusion provoked Hindu nationalist leaders to object directly to the prime minister that the ASI report was blasphemy (Indo-Asian News Service 2007), and they called for a withdrawal of the affidavit and an apology from the ASI (Das 2007). In a few days, the secular coalition government withdrew the affidavit and apologized to the Hindu community—part of a long history of appeasing religious groups. The Union Minister for Law and Justice, H. R. Bharadwaj, "abjectly apologized," stating (as quoted in Bidwai 2007a),

> "Lord Rama is an integral part of Indian culture and ethos . . . and cannot be a matter of debate or litigation. . . . His existence can't be put to the test. . . . The whole world exists because of Rama." He melodramatically added: "Just as the Himalayas are the Himalayas, the Ganga is the Ganga, Rama is Rama. . . . It's a question of faith. There is no requirement of proof to establish something based on faith."

Two ASI senior officials who had authored the affidavit were suspended, and an inquiry was opened.

The opposition grew even as the project continued. Hindu nationalists continued to press on the site's religious nature and its need for protection. After the Apex court ruled that Adam's Bridge should not be damaged in any way, the case ended up in the Indian Supreme Court. Meanwhile, *Images India*, published by the government's National Remote Sensing Agency (NRSA)—part of the Department of Space—claimed that satellite images showed that the bridge might indeed be "man-made" (or, technically, god-made) (N. Menon 2013). Hindu nationalists used this evidence to argue that "science" had finally vindicated their position. In an insightful analysis, Nivedita Menon (2013) notes the central and problematic role of science in the case. Secularists, she argues, took refuge in "science" as the only truth-teller. By opposing the religious nationalists and supporting the canal, secularists put their might behind both science and a development agenda, although the two are not necessarily related. Menon argues that secularists thus conflated science and development with the modernist project. In a strong opinion column, journalist Praful Bidwai (2007a) chastised the government both for appeasing religious groups and for not using the power of science as the final arbiter of truth. He argues that not using scientific evidence would be "giving in to the idea that faith must always trump history, archaeology, even geology—which explains the existence of natural formations like Adam's Bridge—and accepting that the project must be scrapped because of myths and scriptures, not fact" (Bidwai 2007b). Further evidence, which came from the Geological Survey of India, argued, like the ASI report, that there was "no evidence" of man-made structures and that geological analyses of the reef revealed a natural formation from thousands of years before humans settled in the region (Bidwai 2007b; N. Menon 2013).

Alongside the debate on secularism and religion, environmentalists were a second set of detractors of the SSCP (Jaffrelot 2009; N. Menon 2013; Feagans 2014), motivated by two main arguments (Ramesh 2004). First, they asserted that the canal would damage a "delicate" ecosystem and impact many marine species that were unique to the region. Second, they contended that the reef provided a barrier to tsunamis and that dredging the canal would increase the risk of the damage from future tsunamis. Environmental activists and biologists highlighted the potential loss of "native" species and described some of the "richest" ecosystems in the world filled with "the most important biodiversity areas of mainland India": "unique" species of marine plants and animals, including coral reefs, "lush" mangroves, fishes and seahorses, and algae and other marine plants (Kathal 2005; Ramesh 2005; Rodrigues 2007; Rao et al. 2008). *Local* and *native* are important markers here since they highlight that conservation must happen at a local level and, as a result, efforts must focus on the "natives" that do not exist anywhere else in the world. Ecosystems are seen as "delicate," "fragile," and "closely interconnected," involving organisms that are "interdependent" on one another in a harmonious and ecologically "balanced" system of biota and biodiversity (Stanley 2004; Kathal 2005; Rodrigues 2007; Rao et al. 2008; Feagans 2014). In this case, they argued that the region had the second highest level of marine biodiversity on earth (Kathal 2005). Any damage would shift the ecological balance of a delicate, interdependent ecosystem and have a "devastating" impact, creating a cascading effect of ecological imbalance and resulting in endangered and extinct species. As always in such environmental debates around development, there are officials who claim that environmentalists are overstating the negative impact to the ecology of the region (*Hindu* 2005).

As the drama unfolded, we begin to see a convergence in the two lines of argument—religious and environmental. First, religious nationalists began highlighting the negative environmental impact in their opposition. Second, environmentalists began to use tactics similar to those of religious nationalists by drawing on the "sacred" to describe the environment (Feagans 2014). Promoting the religious dimensions of the case builds on a long legacy of environmental perspectives in India, such as the idea of "sacred groves" (Gadgil and Vartak 1974; Malhotra et al. 2007; Apffell-Marglin and Parajuli 2000; Kent 2013), which claim a unique relationship between Indians/Hindus and their local environments. Some suggest that sacred groves are part of "an ancient tradition" in India, predating agriculture and representing a valuable legacy of nature conservation, in which individual species—and at times entire biological communities—are protected "because of their association with some deity" (Gadgil and Vartak 1974; Gold and Gujar 1989). Thus, "sacred groves" come to represent both a botanical and a cultural heritage. These tracts, scattered throughout India, have been sites of environmental conservation and management. In contrast to Western modes of exploitation, some activists and scholars have highlighted the spiritual and religious relationship of locals with their surroundings.

In contrast, feminist and environmental scholars have critiqued any simplistic idea of sacred groves (Kent 2010, 2013). Sumi Krishna (1996) argues that such romantic representations betray the nativism as well as the caste and gender politics of such spaces. Romantic descriptions of sacred groves ignore how such groves are governed. Brahminical patriarchy shaped by strict gender and caste segregation often restricts entry. As a result, women and those of lower castes are often prohibited from entering groves even when the associated deity is female. Sacred groves are also diverse sites, some quite recent in origin and individually owned rather than communal. We return to the relationship of environmentalism to nativist, caste, and gender politics later in this chapter.

Despite these challenges, there exists a strong strain of Indian environmentalism both within India and globally that is deeply rooted in the celebration of the religious and sacred (Gadgil and Vartak 1974; Shiva 1988; Jain 2011). This strain of environmentalism, called the "new traditionalist" discourse, asserts that "'traditional agriculture' was ecologically balanced, and was practised by self-contained communities, . . . and that women, forest dwellers and peasants were primarily the keepers of special conservationist ethic" (Sinha, Gururani, and Greenberg 1997). In short, there remains a vibrant debate on whether such forest communities should be seen as "relic" populations in harmony with their surroundings and removed from outside contact or as populations integrated with larger areas of politics and culture (Karlsson 2006). We return to challenges of these claims in the next section.

Environmentalists drew on the legacy of environmentalism and religion to highlight the "sacred" aspects of the region. Soon this campaign went international. In 2005 UNESCO declared the *Ramlila*, the traditional performance of the *Ramayana*, an element of intangible cultural heritage. A push followed in 2007 to declare the Rama Setu a world heritage site or a "sacred site" (Narain 2008; *Times of India* 2009). Activists organized an international meeting at the Linnaean Society in London in 2008 to save the Gulf of Mannar from destruction by the SSCP (Momin 2008). Since only the Indian government can make such a declaration, more than ten Hindu umbrella organizations around the world joined to launch the Ram Setu Campaign (www.ramSetu.org) and petitioned the government; they received no response.

From September 2007 until April 2010, Hindu nationalist organizations led a well-orchestrated campaign (van Dijk and Mamadouh 2011), mobilizing their base to launch public protests and hunger strikes. A strong police response, including tear gas on some occasions, added to the public disruptions (*Times of India* 2007). The hard-liners accused the government of "hurting Hindu sentiments" by suggesting that Hindu gods were merely mythological figures (Madhur Singh 2007). Jaffrelot (2008) argues that the form of outrage that Hindu nationalists have developed rests on a discourse of victimization that has historical roots. A "vulnerability syndrome," he argues, was shaped in the late nineteenth century as a reaction to the perceived erosion of powers of Hindus in favor of

Muslim rulers and European colonialism. The majoritarian complex of inferiority translated into a vigilance about sacred identity symbols under attack. What is interesting in this instance is that the aggressors are other Hindus and the state (Jaffrelot 2008).

As Nivedita Menon (2013) argues, what is striking about the case is that while the Indian state invoked secularism and "science" to support the canal Hindu nationalists, far from invoking faith and belief, *also* claimed scientific evidence to suggest the bridge was made. The Supreme Court asked the central government to investigate alternate plans for the proposed canal to see if one could be found that would not damage Rama Setu. The government set up a committee headed by engineer and environmental scientist R. K. Pachauri in July 2008. The committee's report, filed in 2009, opposed the SSCP but shifted the terms of the debate to emphasize only the ecological consequences of the project (*Hindu* 2013a). Ultimately, in 2010 the project was suspended, and it has since remained dormant. The issue resurfaced again during the election in 2014 (Narasimhan 2014; Scott 2014). Whether the case is taken up again remains to be seen.

To understand postcolonial environmental politics in India, one needs to recognize the relationship between science and religion in India. As we have seen in previous chapters, while science and religion are intimately and inextricably interconnected in India, this association has its own set of exclusionary politics of gender, caste, and class. Religion has a complex and varied history in India. Despite the presence of goddesses and female power, Hinduism is a deeply patriarchal religion. Hindu nationalism, in both its violent, strident incarnation and its softer form of a "new age/ *Hindutva lite*," has ushered in a renewed pride in the nation's "Hindu" heritage, its success in a globalized world, and the emergence of a large and vibrant middle "consumer" class (van der Veer 1994; Simanti Dasgupta 2015). This pride is amply in evidence in postcolonial environmentalism as India reclaims its "unique" heritage and contrasts its differences from the West. It is clear that religion has never gone away; rather, it has been transformed into new, secular practices—including scientific practices. Narratives of rescue and return are shared by both Hindu nationalists and environmentalists.

Science only becomes a subject in opposition to nonscience (or pseudoscience, or religion). It's not just that "science" is heterogeneous but that it is relational and processual, like gender.[2] Postcolonial studies remind us that colonized nations and natives were never passive victims in the story of colonialism but actors with agency, engagement, and resistance. The resurgence of Hindu nationalism, and its refiguring of Hinduism as a global and scientific force, is a testament to such resistance.

ACT 2. DIVIDING UP THE EARTH: CASTE, GENDER, AND SUSTAINABILITY

The field of religious studies has traced the deep roots of many religions, especially animist religions, to the personification of nature and the divinity of trees, forests, and their inhabitants. Yet, as Pramod Parajuli (2001) reminds us, there is a difference between religious texts and ecological accounts of lived experiences that are not filtered through such texts. Lived experiences point us toward more grounded histories of "ecological ethnicities" that reveal how social hierarchies of gender, caste, and class can structure everyday living and ecological considerations.

Many scholars have documented how Hindu religious rituals have often embedded local ecologies into their ritual practices (Jain 2011). I am also interested in the converse, that is, how scientific theories have embedded religious and local practices as models of environmental sustainable management (Whitmarsh and Roberts 2016). In particular, I have been fascinated by the copious publications of well-regarded Indian ecologist Madhav Gadgil, who, along with other Indian and international collaborators, has since the 1980s proposed and elaborated a theory of sustainable resource partitioning based on caste.[3] In examining the myriad local ecological practices that span India, these ecologists (and indeed many others) support a long-standing claim that one way in which the indigenous is marked from the modern is that the indigenous does not make a distinction between objective and subjective and between scientific and social knowledge. By extension, they suggest that science and modernity do distinguish between the objective and subjective knowledge. Their theories of resource partitioning reveal the ways in which the celebration of the native and indigenous can shape scientific theories. A powerful strand of Indian environmentalism has long celebrated the "indigenous" as more ecologically responsible and sustainable than the modern and has placed India's ecological crises squarely at the feet of Western modernity (Shiva 1988, 1993, 2016; Nandy 1988; Mies and Shiva 1993; Viswanathan 1997). There is a seductive logic of the "indigenous" as pure, natural, and desirable. The seduction works effortlessly alongside orientalist ideas of the "East" as regions that are closer to nature than the West. Also evident is the strong criticism these theories have evoked.

The scholarship on caste has ranged from those who focus centrally on its ritual-religious dimensions within Hinduism to those who stress its political-economic ones (Ghurye 1969; Bayly 2001; Dirks 2001; S. Guha 2013). Whatever one's genealogy of caste, there is little doubt that caste emerged as a potent social, political, and economic system in colonial and postcolonial India. Caste groups are further splintered into numerous subcastes. As populations grew, the caste system became more intricate and elaborate, rigidly defined and perpetuated by divisions of labor and profession on the one hand and religious and commensal practices and marriage rules (endogamy) on the other. Estimates vary, but rough estimates are that the main caste groups are divided into three thousand castes and twenty-five thousand subcastes, each based on specific occupations (*BBC News* 2017). Indeed, segregation along caste lines has been a central and important organizing practice in both historical and contemporary India. Even after religious conversions out of Hinduism, caste identity follows individuals and continues to shape their privilege and lived experiences. Caste and its social manifestations have shifted and transformed with time, making it difficult to essentialize a singular caste system across region and generations (Quigley 2000) and leading some to describe caste as a system of many avatars (Srinivas 2000).

It is this social segregation of caste that inspired ecologist Madhav Gadgil and his colleagues to elaborate an ecological theory of caste. Since the early twentieth century, theories of resource partitioning and niche partitioning have been important in the field of ecology. Such theories attempt to explain how natural selection enables diverse species to coexist in shared habitats. According to the theories, species can coexist when organisms partition the resources (space, nutrients, and other biotic and abiotic resources) rather than competing for them. This partitioning is enabled through specialization and niche differentiation such as niche segregation, niche separation, and niche partitioning. As sustainability rose in importance in the ecological literature in the 1980s, Gadgil and other ecologists in India and abroad proposed, in a series of empirical and theoretical works, that societal segregation around lines of caste has served as a mechanism of ecological resource partitioning, leading to ecologically more sustainable and diverse habitats in India (see, for example, Gadgil 1987). They argue that the caste system's origins can be traced as a response to resource scarcity—a response that evolved as a social mechanism and adaptation for ecological survival—and that the evolution of caste has given rise to a sustainable ecological system of resource use. In this theory, the caste system was devised as a "unique system of cultural adaptation to the natural environment" in response to a resource crunch. The social system "crystallised in the form of caste society, defined by its hereditarily prescribed modes of subsistence" (Gadgil and Guha 1992: 103). Further, they argue that the caste system

> on the one hand, forced its members to share natural resources, and on the other, created the right social milieu in which sustainable patterns of resource use were encouraged to emerge. . . . It was a social system which both forced and cajoled the social being right from birth to adopt sustainable cultural mores.
>
> Within the caste system, birth determined a human being's occupation. An "ecological space" and its natural resources could only be used by a definite occupational group. This "resource partitioning" helped to reduce competition and, hence, conflict among human being over scarce natural resources, and to create the right psychological environment: the allottees of an ecological space developed sustainable patterns because they had no worry that their resources would be snatched away from them and probably also because they knew that if they exhausted the resources in their own space they would not be allowed to use any other.

Indeed, in this reading, one can see how "caste" becomes a very Foucauldian biopolitical regime in organizing life and resources. It has profound implications in terms of which lives are deemed worthy of highly valued resources and which are relegated to more precarious living. As always, because history is written by the powerful, such differences are often elided, and within the ecological literature, this theory of caste emerges as an important insight in the literature in resource ecology and ecological sustainability. For example, in a review essay discussing the importance of "the basic ecological unit," Berkes, Kislalioglu, Folke, and Gadgil (1998) argue that "traditional knowledge may complement scientific knowledge by providing practical experience in living within ecosystems and responding to ecosystem change." The claims by Gadgil and his various colleagues have with time come to normalize the idea that social organization such as the one in India can lead to more sustainable ecologies, with little mention of the oppressive social contexts in which they emerged and are enacted (Dupré 2002; Richerson and Boyd 2005; Pirta 2009; Zhang 2010; Museka and Madondo 2012).

Their theories have not escaped the eyes of their critics. For example, Sumi Krishna (1996), in a wonderful analysis of theories of resource partitioning around caste, questions and refutes their claims that the origins of caste have an ecological basis. She also powerfully critiques any representations of caste as a benign form of resource partitioning, arguing that doing so ignores caste's oppressive and violent histories. It is not as though resources were equally divided among castes; rather, the privileged claimed the lion's share of the resources while others were forced to make do with very little. History has well established that resources were protected, often through intimidation and force. These practices were contested, and indeed resisted. As Sumi Krishna also points out, lived realities of caste remind us about the centrality of women and gender roles in resource partitioning. Endogamous marriage rules have always rendered women as objects of economic exchange. Ecological lives of families and communities are practiced literally on the empty stomachs or the burdened backs of women. The high ideal of sustainability in these contexts is often achieved through exploiting the labors and bodies of women.

A few years later, Gadgil and Guha (1995) responded:

> We were charged with promoting the caste system as an "eco-friendly" alternative to present patterns of resource use. We were even accused of a Brahmanical bias. This, we submit, is a criticism

that owes everything to political correctness and nothing to reason or accuracy. In prefacing our account of caste as a system of resource use, we state categorically that caste society "was a sharply stratified society, with the terms of exchange between different caste groups weighted strongly in favour of the higher status castes." Perhaps we should have italicized these words, or have had them set in bold type. Not that it would have helped; in India's politicized climate today, willful misinterpretations of scholarly work are nearly inevitable.

Some three decades later, I am less interested in whether Gadgil, Guha, Malhotra, and their various colleagues inside and outside of India are casteist, or whether they intentionally valorized the caste system. Rather, I am fascinated by how their work has been taken up—how have scholars in ecology, environmental, and development studies understood this work? As I quickly discovered, their work, prolific to begin with, has had wide and deep impact. Each of their papers has been cited repeatedly, often hundreds of times. In tracing their theoretical impact, several patterns emerge.

What is striking is that caste repeatedly is represented and highlighted as an "innovative" social system in India, with its religious inflections and connections downplayed. There is virtually no mention of religion in this history. Despite the authors' protestations, they explicitly present caste system as "adaptive." For example, one of the early and influential papers by Gadgil and Malhotra, published in the *Annals of Human Biology* in 1983, is titled "Adaptive Significance of the Indian Caste System: An Ecological Perspective." The very idea of an "adaptation" suggests that it enhances the evolutionary fitness of an organism—and in this case a diversity of humans. Nothing in their papers suggested that the caste system was maladaptive for some members, just that the whole system worked sustainably.

The legacy of their work has become a "frozen moment" in the sustainability literature (Haraway 1991). Caste is cited again and again as a basic ecological unit and their as work as an account of how this model of social organization is effective. While Gadgil and his colleagues may have intended to present a more nuanced argument, three decades later their work has been fossilized as a citation needing no comment—the literature now presents Gadgil and his colleagues as having demonstrated that caste is a magnificent and unique example of social organization and social ecology. For example, a recent piece by Nakahashi, Wakano, and Henrich (2012: 414) cites Gadgil's work as evidence to argue, "In more complex human societies, occupational specializations of the kind associated with complementary interactions emerge principally in relations *among* social groups, with whole groups, castes, classes, or guilds specializing in one or another skill." In discussing the evidence for tribal instinct evolution, Peter Richerson and Robert Boyd (2005: 271) use Gadgil and Malhotra's (1983) work to support the claim that "In civilian life, symbolically marked units include regions, tribal institutions, ethnic diasporas, castes, large economic enterprises, religions, civic organizations, and, of course, universities."

While Gadgil and his colleagues acknowledge shifts and changes in the caste system, their work is premised on two main methodological claims: First, that by examining extant populations (which is what their theories exclusively rely on—the large amount of anthropological data on the peoples of India), we can surmise earlier caste organization. Because multiple caste groups currently share the same geographic area without competition, they see this as evidence of the culmination of an adaptive paradigm that evolved over centuries. Second, that the problems with caste oppression today are blamed on the ills of colonialism, or more recently on neoliberalism (Gadgil and Guha 1992), which have corrupted earlier sustainable practices of caste. At the end, what emerges is a nativist and nostalgic view of ancient India, and an interpretation of caste as a primal, socially innovative, and equitable system of resource partitioning. This nativism mingles with orientalist thinking as caste is valorized abroad.

In contrast, the history and sociology of caste presents a much more complex picture—both in the many evolutions of castes but also in their fluidity and mobility in lived experiences. As Dipankar Gupta (2000: 1-3) states,

> It is more realistic to say that there are probably as many hierarchies as there are castes in India. To believe that there is a single caste order to which every caste, from Brahman to untouchable, acquiesce ideologically, is a gross misreading of facts on the ground. . . .

As the "book view" of the caste system is derived largely from sacerdotal Hindu texts, members of the upper castes find it extremely agreeable . . .

Nevertheless, the difference between the book view and what happens on the ground is quite remarkable and must be attended to.

The Brahminical dictum of the established four castes that has disciplined all Indians into submission, Gupta argues, is patently false. Instead, he argues, "throughout history there have been caste revolts and caste mobility." Caste is a more fluid system than popular and even academic mythography would have us believe.

The caste system has become biologized and ecologized as a discrete and homogeneous unit, quite opposite to the lived realities that D. Gupta (2000) describes above. For example, John Dupré (2002: 131), in discussing human evolution as a cladogenetic process involving the generation of many different cultural species, considers the significance of treating distinct cultures a genuine taxonomic entities, what he calls "cultural species." In a footnote he cites Gadgil and Malhotra to say that "a specific application of such an attempt to the Indian caste system has been attempted by Gadgil and Malhotra (1983), though they concentrate almost exclusively on ecological factors."

There is little mention that this sustainable system has been possible only through technologies of power. Caste has been stripped of any references to oppression or violence. It is important to note that theories that elide and erase the lived realities of caste are easily appropriated by Hindu nationalism and its recovery and celebration of a grand precolonial Vedic period. Williams and Mawdsley (2006) point out:

In an earlier piece on "The Ecological Significance of Caste" . . . , Madhav Gadgil comes close to justifying caste division on the grounds that "traditionally", differential caste taboos over natural resource use partitioned these resources, placed castes into different "trophic layers" and so ensured ecological sustainability. This normalizing of highly unequal social divisions through the metaphors of natural science is one clear example of the dangers of evoking a precolonial golden age.

Representing the caste system, in its origins and evolutions, as benevolent elides its oppressive history. While indeed Gadgil and his colleagues may not justify the caste system and may acknowledge its oppressive nature in contemporary India, their work's influence draws on nativist and orientalist ideologies that see societies in the "East" as more primitive, less modern, and more at one with nature. Such wide-reaching orientalist discourses complement nativist discourses that essentialize and romanticize indigenous, local cultures (Mawdsley 2006). Ecological theories of resource partitioning valorize caste politics as important indigenous systems that sustain diversity and ecological sustainability while ignoring its histories of violence and inequality.

ACT 3. SAFFRON OR GREEN? ECOLOGICAL THEATER IN POSTCOLONIAL INDIAN ENVIRONMENTALISM

> Indian philosophy on the contrary is that you can enjoy life only when the surroundings are healthy and harmonious. That is why in our mythology trees are worshipped. Even water, earth and sky are worshipped. We have been taught to replenish what we take from the Nature. India's ancient message of sustainability thus can show the way to the fight against Global Warming and Climate Change.
>
> PRIME MINISTER NARENDRA MODI

Secular environmental movements have an ongoing, albeit ambivalent, relationship with Hindu nationalism. In summarizing the various factions of environmental activism in India, Sivaramakrishnan (2011: 106) writes, "Varieties of nationalism, parochialism and inter-community hostility have apparently fed on varieties of environmentalism that have informed or drawn upon what has been described by these commentators as identity politics." The SSCP (Sethusamudram Shipping Canal Project) demonstrates the theater of such postcolonial contestations, where multiple and opposing constituents claim their rights. While the SSCP began during a Hindu nationalist coalition government, when construction began the same groups opposed it on religious grounds. We also saw the growing synergy between science and religion as environmental discourse embraced the sacred and religious discourse embraced the environmental. While religious nationalists opposed the construction of the SSCP and have opposed other development projects (M. Sharma 2009, 2011), they are not antiscience or antitechnology. Indeed, alongside such contestations we have seen the wide embrace of the market, science, globalization, consumerism, and neoliberalism in new formations of environmentalism within modern Hinduism. Similarly, in the ecological literature we see how the oppressive histories of caste get elided as they move into more "objective" scientific terrain; caste transforms into ideas of exalted "Indian culture," celebrated by scientific and religious sources.

In his book *Green and Saffron*, Mukul Sharma (2011) argues that "a large part of the panoply of environmental politics in India today in fact reveals some political allegiances or affinities with Hindu nationalist and authoritarian forces." *Green postcolonialism* refers to the growing body of work on environmentalism in the postcolony (Huggan and Tiffin 2008). *Saffron postcolonialism* refers to the environmentalism emerging from Hindu nationalists in India (saffron is associated with the sacred in Hinduism). As the SSCP project demonstrates, there is a deep resonance between the green and saffron. But are they the same? Using several case studies, M. Sharma (2011) contends that modern Hindu environmental politics glorifies the landscapes of India as Hindu and romanticizes an ancient philosophy of nature while simultaneously condemning modernization, Westernization, and globalization. He argues that an authoritarian brand of activism attempts to institutionalize "eco-naturalism" and "eco-primitivism" and to celebrate the hierarchical, unequal order of traditional Indian villages, and he cautions us about the greening of Hindu nationalism. Amita Baviskar (2012) in her review argues that Sharma concedes too much to Hindu fundamentalism. Arguing that many everyday practices around rivers, mountains, and forests, like the river Narmada being worshipped as "Mother Narmada," should not automatically be written off as a regressive politics of Hindu nationalism. She calls for a deeper understanding of religion and politics that can celebrate everyday environmentalism without rejecting it as part of a violent Hindu nationalism.

In a contrast to Sharma's analysis, Pankaj Jain (2011) suggests that we should learn to distinguish hierarchical religious institutions and their structures from local everyday spiritual practices of communities—that is, we should move from an institutionalized Hindu religion to talking about lived philosophies of dharma. Similarly, Emma Tomalin (2009) distinguishes "bio-divinity" in religious traditions such as Hinduism from "religious environmentalism"; the latter, she argues, is a product of a post-materialist philosophy emerging from the West. In her view, such a distinction allows us to separate environmentally sustainable practices from the politics of a virulent Hindu nationalism. She also cautions against embracing the language of environmentalism since in many cases biodivinity is centrally about religious ritual practice, and any impact on the environment is a byproduct rather than the main focus.

One of the biggest takeaways in examinations of postcolonial environmentalism in India is that any easy separation of religious institutions and local spiritualism, or of local practices and religious orthodoxy, is difficult (Baviskar 2007). Much of the analysis of a canonical religion or spirituality ignores questions of gender or caste, both central forces that shape hierarchies and experiences. Local spiritual and cultural relationships with the environment often span religions and may not always be associated only with Hinduism. In a context of politicized religious conversions and reconversions that have riddled India for several centuries, there remain relatively few spaces of "pure" religion or spirituality. Religion is no longer a private or local space removed from the virulent politics of Hindu nationalism. No community has remained untouched from the politics of colonialism, gender, caste, and religion. The varied and multiple narratives of religious experience in a country like India, where Hinduism and indeed all religions are deeply entrenched in social hierarchies, reflect centuries of the entanglements of the natural and cultural.

Looking at the landscape of postcolonial environmentalism, we see that science and religion, the local and transnational, and religion and spirituality are thoroughly intertwined. The rise of an outsourced, privatized, and neoliberal mode of governance indicates a "cosy relationship between business and political interests, supported by a compliant administration" (S. Sarkar 2018). Indeed, India has the "dubious distinction" of having more environmental conflicts than any other country in the world (S. Sarkar 2018). The privatized model of government is ripe for an increasingly corporatized industry of Indian gurus and swamis. The many gurus, sadhus, temples, and ashrams that have emerged in recent years and run highly profitable enterprises that commodify and sell spirituality. There is no business in contemporary India like the god business (Kumar and Lokhande 2013; Kala 2014). While some enterprises celebrate spiritual and local practices of dharma or biodivinity as environmentalist, Hindu nationalism has also ushered in more familiar forms of Western environmentalism. Thoroughly "modern" gurus run global enterprises that have embraced "green" environmentalism. For example, Sri Sri Ravi Shankar, the famous spiritual leader and founder of the Art of Living Foundation, has developed innovative farming practices. Through the Sri Sri Mobile Agricultural Institute, farmers are encouraged to embrace biodynamic, organic farming techniques and soil and water conservation methods. While often touted as secular, this thoroughly modern institute is one of the newest global Hindu movements (Jain 2011). Drawing on the allure of Indian spirituality, Art of Living centers advertise their services for the ailments of modernity through "classes in yoga, meditation, breathing exercises and spiritual wisdom." Sri Sri Ravi Shankar's rhythmic breathing exercises, called "Sudarshan Kriya," are touted as the concentrated wisdom of the ancients yet have been patented (Fish 2006). The Art of Living Foundation has collaborated with scientists to test and prove the effectiveness of Sudarshan Kriya for calming and reducing stress (Brown and Gerbarg 2005). This easy and effortless melding of Hinduism and science is pervasive in contemporary India. Others also bring an ecological bent to their spiritualism. Amma (Mata Amritanandamyi), the "hugging saint," has established GreenFriends, an ecological revival in greening the environment through large-scale forestation efforts and wide-scale use of renewable energy (see Amma n.d.). In concert, she also promotes "eco-meditation," a melding of ecology, nature, and the human mind (Jain 2011). Similarly, the Sathya Sai Baba Organization in the United States launched the program "The Earth—Help Ever, Hurt Never" in 1995, promoting reusing and recycling materials such as batteries, junk mail, shopping bags, greeting cards, and shoes (Jain 2011). A perusal of India today reveals a country teeming with such initiatives—initiatives that embrace science and Hinduism.

While such versions of modern environmental Hinduism embrace capitalist and neoliberal practices, there is little transformation of the actual customs and ideologies of Hinduism. Some activists have pointed out that while Prime Minister Modi has begun a campaign of Swachh Bharat, or Clean India, religious rituals, especially the dispersal of ashes after cremations, are among the biggest polluters of the holy rivers (Abdi 2015). Similar environmental problems have also emerged within diasporic communities abroad, such as in the growing Hindu population in New York, who are accused of polluting local rivers (Dolnick 2011). While many practitioners have individually embraced electric crematoriums, there has been little reform of Hindu customs to make them more environmentally friendly. In contrast, we have seen widespread celebration of religious institutions embracing solar-powered roofs, technologically sophisticated irrigation systems, and water treatment plants. The rhetoric of "green" in modern Hinduism is centrally about Hinduism rather than environmentalism. The "indigenous" India is rendered a fetishized object, a benign premodern site of nostalgic return. The social stratification of Hindu nationalism and the increasingly neglected sites of rural life are lost in the nostalgic rhetoric of return to an idyllic past. The nostalgia resides in empty celebrations of the local and indigenous in global circuits, even while these very sites remain embattled within a global and neoliberal India.

But as Deepika Bahiri (2018) cautions, we should not read the proliferation of these environmentalist consumerism as mimicry of "Western" form. Rather, "recombinant hybridity reconciles both native and global forms of elite existence. . . . In the high noon of globalization, the privileged postcolonial subject binds his or her native prejudices, hierarchies, and ways of living with norms evolving in the new world order" (Bahiri 2018: 136). One of the striking aspects of debates on postcolonial environmentalism across the globe is the contestation around modernity, indigeneity, and what constitutes "nature." Alternate models of biodivinity have much in common with strands of ecofeminism. Both offer critiques of Western science and Judeo-Christian philosophies that shape modern living. They are often grounded in essentialist ideas about women and indigenous groups being closer to nature. Both share a nostalgia for the nonmodern or premodern and look for a return to "real" or primal nature. Both pay deep and careful attention to social practices that shape the ecologies and environments around us. Yet in the celebration of authentic, native, or essential nature, the politics of the "natural" get lost. De la Cadena (2010: 360) persuasively argues that our only hope beyond these binaries is our willingness to give up "old answers (and fears) that mirror each other." She insists that we can neither demonize the indigenous as traditional, archaic, and dangerously prone to antidemocratic fundamentalism, *nor* valorize it as always inherently good. We must attend to the binaries that continue to haunt our theories and politics.

UNBRAIDING HISTORIES: BEYOND THE NOSTALGIC POLITICS OF RETURN AND RESCUE

Stuart Hall (1996) has argued that one of the achievements of postcolonial studies is that it has made visible the long-enduring legacies of colonialism; we increasingly see that the *post* in *postcolonial* is "under erasure." Just as Hindu nationalism's politics of erasure and victimhood (Jaffrelot 2008) traces back to its colonial roots, so too does the history of environmentalism. Grove's (1995) history of ecological thought demonstrates that modern environmentalism and its critical focus on the human originates in colonial rule. The state's long-term security was dependent on the colonies, and so using land and environmental management to deal with ecologically destructive commercial expansion was in fact easier in the colonies than on the mainland (Grove 1995). There is also something deeply paradoxical about the nostalgia for a "native" world of the past, as though ecologies are static and unevolving. In fact, Albert Crosby's *Ecological Imperialism* (2004) ushered in a vast body of literature that chronicles how profound the ecological impact of colonialism was. The "portmanteau" of biota (plants, animals, and pathogens) found in or transplanted to the colonies enabled the expansion of Europe and radically transformed the globe (Crosby 2004). In recognizing the deep interconnections between the human and nonhuman, between nature and culture, we now recognize how central ecological landscapes have been in human history, and vice versa. Colonial landscapes were profoundly transformed, and a politics of rescue grounded in reverting a postcolonial world to a precolonial one ignores the complexities of colonial ecologies.

Sivaramakrishnan (2011), surveying the loaded politics of nationalism and the environment in postcolonial India, suggests that we should distinguish between two kinds of nationalism—thin and thick. Thin nationalism is environmental politics "where affinity to a collective called the nation is evoked for purposes of expressing ethnic affiliation, or empty ritual serving political endgames" (Sivaramakrishnan 2011: 87). In contrast, thick nationalism is "a positive and capacious sentiment that inspires people moved by it to strive for exalted values and ethical conduct . . . a robust and inclusive nationalism can find secular reasons for amity, cooperation and the pursuit of human welfare across the widest range of the citizenry and promote vigorous debate on human values that must be accommodated in national priorities" (Sivaramakrishnan 2011: 87). What has been apparent in the rise of Hindu nationalism is that key elements—the SSCP debate, theories of resource partitioning, or the revival of nostalgia through a politics of return to a past "golden" age—are examples of a "thin nationalism."

Within such naturecultural histories and the need for rescue of the present, the unquestioning desire to return to a "native" past and the nostalgia about a world long gone seem entirely misplaced. Given the profound environmental changes that colonialism has wrought, such a return, even if desirable, is not possible. We can understand such desires only in the context of Said's opening epigraph. Indeed, the colonized can only return to their land through a politics of imagination, and such nostalgia is a politics of recovery (DeLoughrey and Handley 2011). Science and Hinduism are global institutions with intertwined and complex histories, shaped through larger historical and geopolitical forces. We need to attend to both the universal and the particular in understanding these new spaces of postcolonial environmentalisms.

In recognizing the impossibility of "pure" politics, postcolonial environmentalism begins to reimagine worlds that eschew the easy binaries of science/religion, nature/culture, sacred/profane, and human/nonhuman. Yet the debates and discussions in postcolonial environmentalism show the tensions precisely around such binaries. We are only beginning to recognize the profound ways in which nonhuman worlds have shaped human history. Rewriting the history of the colonial and postcolonial through a non-human-centric analysis awaits new insights. If we want to engage in a historical model of ecology, we must enter a "profound dialogue with the landscape" and undo our epistemological tendencies to decouple nature and history, human and nonhuman (W. Harris 1990: 175). Such reductionist and decoupled histories have helped mystify colonialisms and the histories of forced migrations, suffering, and human violence (DeLoughrey and Handley 2011). As Haraway challenges us, we must develop new accounts of the world that dismantle nature's constitution as "other" in the histories of colonialism, sexism, racism, and class domination. We must "find another relationship to nature besides reification and possession" (Haraway 2007: 158). And while indigenous practices may offer such alternate visions, we must be mindful about the ways in which there is no place outside of politics. The messy histories of gender, race, caste, and religion create contradictory spaces of possibilities. But it is within the ruptures of this messiness that we need to create feminist spaces of imagination, spaces not locked in rescue and return but willing to develop Harris's "profound dialogue with the landscape" to see new processes of naturecultures emerge. For example, while some may reject the mainstream retelling of the *Ramayana* on the grounds that it is authoritarian, masculinist, and racist, we can still be open to the epic's imaginative possibilities of an agentic natural world, one where forests are not "resources" but rather speak back to humans. In Ruth Ozeki's *A Tale for the Time Being,* one of her characters, Oliver, creates a NeoEocene, anticipating the future environment and planting for that future. Oliver is not locked in a nostalgia for the past or delusion about the present and its capacity for stasis, but rather he opens up the future through a dialogue with the landscape, crafting new futures and waiting for the landscape to speak back. And as he remarks, such dialogue requires patience. After all, we are "short-lived mammal[s], scurrying in and out amid the roots of the giants" (Ozeki 2013: 61). In such patience and humility lies the possibility that we may escape the many afterlives of colonialism and imagine new incarnations and landscapes of justice and freedom.

AVATAR #4

The Story of Néram

A new iteration, #1729, was unfolding on Kari. Néram, the node, or Kanu, of time, long considered the jokester of Avatara Lokam, basked in the joys of new beginnings. These moments were some of the most creative and inventive. Néram was illusory, moving effortlessly across time. If you caught a glimpse of it, all you would see would be a bright flash or a spark, and before you knew it, space-time continuums could collapse, or fold into each other, or smooth out. Néram's humor was not always appreciated. Sometimes in one quick instant it could undo the work of Uruvam, Amudha, and Nādu, much to their annoyance. But after flexing its temporal muscles, Néram would laugh, right the wrong, and return Earth to its planetary tempo. With a new evolutionary cycle well under way, Néram was excited by the emergence of a new humanlike species, *Homo impuritus*, along with a second species that resembled the earlier *Homo sapiens*; both inhabited the planet together, with watchful eyes on each other. Some individuals became friendly across species, creating hybrid communities. Néram was clever and entertaining, with a penchant for whimsy, and delighted in intrigue and surprise. This time around, Néram had worked closely with Amudha, and both were pleased with the appearance of *Homo impuritus*.

Néram introduced more obvious clues into its creations, and time became a clear and playful element for the planet. Predictably, some humans invented a clock, ushering in a disciplined, regimented way of life. Néram had introduced multiple temporal schemes into life on Kari—the sense of slow and endless time, the epochal times of geological evolution, and the deep time of the universe. But rather than enjoying Néram's playful registers of time, its unpredictability, its rhythmic meters or its capricious cadences, these humans synchronized all temporal registers into one joyless calendar. An ideological battle emerged. The Time Zealots promoted an absolute clock for the whole world, promising efficiency, synchronicity, uniformity, and world integration. They would eat by the clock, not just when they were hungry; they would sleep and wake up by the clock, not when they were tired or sleepy; they would celebrate and mourn by the clock, not when they were happy or sad.

What surprised Néram was their obsession with origin stories and a resolutely linear view of history. It watched with horror as the Time Zealots rewrote children's history textbooks to present a dry and rote recitation of their version of the world. Everything had a beginning and then moved on in a singularly linear fashion. Néram was most dumbfounded by their obsession with beginnings. The Time Zealots introduced the idea of celebration time—their followers celebrated the hour of their birth, objects were known by the year of their discovery or invention, national histories were marked by moments of independence. The dawn of each year was marked by a celebration—"Ring out the old, and ring in the new," they chanted each year. Their celebratory scheme was coupled with an obsession with reproduction. They seemed intent on listing generation after generation—parents, grandparents, great-grandparents, great-great-grandparents, and so on—as though it were some meaningful exercise. Amudha had created blood to transport oxygen to the cells, but humans had turned blood into some mystical element of historical truthtelling! Some nations had even made blood lineage, or the principle of *jus sanguinis*, the criterion for national belonging.

Néram had embedded startling clues in physics and astronomy, geological surprises, biological leads, and many delightful science fiction stories. There were clues at every turn that time was playful, not a regimen. The Time Zealots ignored Néram's whimsical and playful codes, but the Time Seekers compensated with their enthusiasm. Néram prided itself in having choreographed a rip-roaring and zany production—a theater of life that wandered and meandered through delicious and dramatic moments, some slow, some fast, but altogether completely exciting and enjoyable. The Time Seekers were slowly figuring it out. Given the humans' obsession with origins, Néram anticipated that genetics would become an important tool for human curiosity. What a delight to watch! As Néram's unpredictable genetic cues unfolded, biologists suddenly reclassified organisms based on their genes rather than on how Uruvam had shaped them. So whales and bats were mammals like humans, rather than fish or birds. One fine day, scientists announced that giraffes were actually four species. For *Homo sapiens*, their obsession with tracing genetic pedigrees through reproduction became the sacred test of belonging: who belonged to a particular species, group, nation, or family. What delighted Néram the most was how its ingenuity cleverly dismantled the superiority of some humans in a quick instant. Through genetic tests, white people suddenly discovered they were black or Indian! Indians discovered they were European! Néram had made sure that the markers were distributed haphazardly enough that no coherent narrative would emerge; depending on which set of genetic markers you looked at, you would get a different history or genealogy. But rather than follow their sense of curiosity, *Homo sapiens* transformed genetics into divine revelation. Across the world, they began sending swabs of their cheeks or blood away for analysis. Néram admitted to being surprised that even close-knit families, communities, and nations could be broken apart by one genetic test. How capricious their affections were! With one test, they obsessed about their true "biological" parents and biological families, or the continent of their "true" origin. Their sense of identity increasingly became fantastical, focused on what their genetics told them, not their lived lives!

In contrast, *Homo impuritus* reveled in the genetic results. After the advent of genetic ancestry, they even renamed their species as "impuritus" to signal that they were a mongrel species, with varied and promiscuous pedigrees. They reveled in impurity and openly taunted the inbreeding inclinations of *Homo sapiens* and their penchant for talk of genetic purity. They welcomed those few *Homo sapiens* who shared their philosophy, but as genetic science progressed, the ideological divisions between the two species grew. *Homo sapiens* continued to pursue a regimented, linear, progressive narrative with a singular focus on the future. In contrast, *Homo impuritus* learned to play with time.

Time, in Néram's conceptual universe, was multidimensional. Time could tear, and through those tears you could watch the past or future, or you could travel to and enter these other worlds. There were forever infinite possibilities of the past, present, and future. Yet despite the multidimensionality of time and space, Kari and the lives within it unfolded each moment—one had to live life to experience it. *Homo impuritus* reveled in the cosmic delights of temporality. They watched, they traveled, and they discovered how time unfolded and unfurled. They did not look back or forward but began to learn to live in infinite, cosmic time.

CHAPTER FOUR

Biocitizenship in Neoliberal Times

On the Making of the "Indian" Genome

I suggest that we regard the paradoxes of quantum physics as a metaphor for the unknown infinite possibilities of our own existence. This is poignantly and elegantly expressed in the Vedas: "As is the atom, so is the universe; as is the microcosm, so is the macrocosm; as is the human body, so is the cosmic body; as is the human mind, so is the cosmic mind."

EDWARD FRENKEL, "The Reality of Quantum Weirdness"

Everyone lives in a story, he says, my grandmother, my father, his father, Lenin, Einstein, and lots of other names I hadn't heard of; they all lived in stories, because stories are all there are to live in, it was just a question of which one you chose.

AMITAV GHOSH, *The Shadow Lines*

THIS IS A STORY ABOUT THE STORIES WE TELL ABOUT THE PAST. It is about the veracity of our collective memories, a testament to our contested imaginations of human beginnings and the birth of nations. Lord Brahma, the creator of the universe, it is said, was born out of a lotus emerging from the timeless navel of Lord Vishnu, ushering in the universe of our dreams and nightmares. Or perhaps life began with the Big Bang and the organic "primeval soup" of abiogenesis that simmered in the smoldering cauldron of the young planet Earth. Organic forms began to combine, recombine, and mutate through the ironic plots of hostile worlds, forever adapting and evolving to produce an astonishing array of life across the ages. Perusing the many origin stories of life on Earth, we see the indelible marks of time, whether it is measured by the linear Julian calendar, the macroevolutionary scales of the origin and extinction of species, the microevolutionary scales of neo-Darwinian time, or the cyclical time of Hindu cosmology, where the universe is created, destroyed, and re-created in eternally repetitive cycles of time. We have a recent addition—the timeless matter of DNA. With one swab of the cheek, DNA is said to reveal all at once—our past histories, our present temperaments, and the possibilities of our futures.

Time in its various manifestations has endured the tumultuous lives of histories. It has witnessed the ravages of colonialism and the promises of nationalism. The endless universe is, or perhaps the many simultaneous universes are, caught in cycles of life and death, creation and destruction, birth and rebirth, living out our karma and dharma. How do we sift through these competing origin stories? India presents an interesting case study because in its quest for modernity, the past and the present, science and religion, and modernity and orthodoxy cohere within the political landscape—and we are faced with competing historiographies brimming with mythological, epistemological, and methodological tensions. This is a story about three such tales.

ORIGIN STORIES: NARRATING THE NATION

> Nations, like narratives, lose their origins in the myths of time and only fully encounter their horizons in the mind's eye.
>
> HOMI BHABHA, *Nation and Narration*

"Epistemophilia, the lusty search for knowledge of origins," Donna Haraway (1997: 255) contends, "is everywhere." This is certainly the case in India. The advent of genetic ancestry testing, genetic demographical health analyses such as HapMaps,[1] the 1000 Genomes Project (1KGP Consortium 2015), and personal genome analyses has renewed these obsessions by ushering in an industry, funded by national and international agencies and not-for-profit and for-profit companies, producing new "genome geographies" of India (Fujimura and Rajagopalan 2011; A. Nelson 2016; Reardon 2017). Given the political import of the results, the claims emerging from these studies have been quickly embraced by a wide variety of actors to pursue what often appear to be contradictory political projects.

In her narration of the evolutionary origins of the United States, Oikkonen (2014) argues that national narratives are predicated on the construction of "the nation" and its "people" through a "set of ethnic, cultural, and geographical differences that often exceed national discourse." This discursively produced "nation" and its "people" are critical for producing a coherent narrative of a nation's origin and purpose, one that helps imagine a "national future" in which cultural and ethnic differences are domesticated to produce a coherent whole. The imaginative possibilities of the nation and its future are not limitless but rather predicated on the past. As a result, narratives of the past can become particularly important—and contested—as nations imagine their futures.

A nation like India poses particular problems in creating an uncontested national prehistory. India emerged as a nation only in 1947 and some of its borders with neighbors are still contested. Given how recently the nation emerged amid the violence of the partition of what had been British India, most stories of India's prehistory are necessarily regional and include the histories of its neighbors. The much-celebrated glories of the Vedic period include areas that are now in Pakistan and Afghanistan. The origin stories of the Indian nation constantly elide the tensions of historiography, creating a unique national prehistory in which regional and global histories are deeply entangled. Benedict Anderson (2006: 11–12), in his famous work *Imagined Communities*, argues,

> If nation-states are widely conceded to be "new" and "historical,"
> the nations to which they give political expression always loom out
> of an immemorial past and . . . glide into a limitless future. . . .
>   What I am proposing is that nationalism has to be understood
> by aligning it, not with self-consciously held political ideologies, but
> with the large cultural systems that preceded it, out of which—as
> well as against which—it came into being.

The past is central to India's nationalism and to Hindu nationalism in particular. India's colonial histories with the empires of the Mughals and the British, and with the religions of Islam and Christianity, profoundly shape nationalism's sense of injury and hope. This is the backdrop for understanding the latest round of contestations of India's origin stories emerging out of new data sets roused from the genome. As we shall see, the genome has become an important site of bionationalism as ancient science and medicine are given new life and modern relevance.

In India's archaic modernity, new formations of genomic nationalism are being secured. Here the transnational and global networks of modern science *and* Hindu nationalism (van der Veer 1994), embracing globalization and neoliberalism, have together proven to be potent tools for the consolidation of a modern scientific ethnic bionationalism. As new data, methods, and methodologies emerge, the old, troubled wounds of ancestry are exposed. Molecular analyses are the battleground on which old debates are being waged anew.

The narratives of India's prehistory and the subcontinent's origins make compelling politics. At the heart of these debates are the relationships between West and East and between global and national histories. Are Indians indigenous to India, forming a biologically distinct and unique population? Or are Indians part of a messier, interconnected global story of human migrations? The historical, literary, anthropological, and now genetic evidence is deeply contested. One key and contested story is that of the "Aryan Migration Theory." As with Indian mythologies, many versions of this story have emerged—in the public imagination, among politicians and activists, and among academic scholars in the humanities, the social sciences, and the biological sciences (Subramaniam 2013). Of particular interest are the origins of the Indian population. Are they descendants of an ancient people who arrived on the subcontinent many millennia ago, or do they include descendants of more recent migrations into South Asia?

The most prevalent narrative, and certainly the one I grew up with, was the orientalist version. It went something like this (Subramaniam 2013): The Indian subcontinent was peopled by Dravidians and other aboriginal groups and tribes. The first migration, and the one with the greatest impact, came from the northwest of India as Aryans invaded India around 3000–1000 BCE. With them came Sanskrit, which belongs to the family of Indo-European languages. These Aryans are believed to have ushered in the grand Vedic tradition that produced a well-developed and distinguished scientific, technological, and philosophical tradition. Many other migrations followed: Persians in 500 BCE, Greeks in 150 BCE, Arab traders in 712 CE, Portuguese in 1498 CE. Finally, of course, were the British in 1610, and they ruled the country for more than three centuries.

The Indo-Aryan migration is critical to the orientalist history, which sees the Aryans as the founders of India and thus as having the greatest impact on Indian history. In the south of India, where I spent part of my childhood, these historical origins and the Aryan migration were an important part of how popular culture explained the population's variations in skin color. South India, in particular the state of Tamil Nadu, is populated with people who are distinctly dark, people who are distinctly light, and all possible hues in between. Popular explanations racialized the skin-color differences we saw around us. The Aryans, the story went, were light-skinned people, the Dravidians dark-skinned; and Indians from the north of India are generally considered fairer than those from the south. Thus it was said that the Aryans, a "superior" people, dominated the Dravidians, driving them southward and elsewhere into the various corners and margins of the subcontinent. These groups were seen as the contemporary indigenous peoples of India categorized, for political purposes, as "scheduled" castes and tribes.[2] For someone like me, coming from the south of India and the land of Dravida, speaking the Dravidian language Tamil (which is unrelated to Sanskrit or the Indo-European group of languages), these stories were relevant and deeply personal. They provided the basis of a particular South Indian, specifically "Tamil" authenticity that resisted the dominant North, which, for example, promoted and imposed Sanskrit and Hindi as the national languages of India. The orientalist version also quite skillfully incorporated the caste system and social privilege: upper-caste Tamils have lighter skin because they have some Aryan blood in them; lighter skin equals more Aryan, and therefore those lightest-skinned groups are higher up in the caste system. Since the time of British colonial rule, some scholars have mapped caste hierarchies onto racial hierarchies (Ghurye 1969; Dirks 1996; Viswesaran 2010; Bayly 1995). Indeed, British colonists, in their "divide and conquer" politics, systematically recruited upper-caste Indians into the colonial administration to help govern the rest of the country. Most historians agree that since there often weren't adequate numbers of British in India, the success of the British in India rested on their ability to rule through Indians (B. S. Cohn 1996). Manipulating the complex sociologies of caste and race was part of this strategy, and the long-enduring links between race and caste have never quite gone away (Egorova 2010a; Viswesaran 2010). In a cultural system where skin color has long been strongly correlated with social privilege and remains a durable marker of beauty, desirability, intellect, and social status, these debates were not trivial, and they continue to shape the hierarchical and social stratification of caste.

While there is, in fact, plenty of variation in skin color and bodily features, these mythic stories endure in the public imagination. Yet there is an alternate view of the Aryan Migration Theory, one that dismisses the theory as a Western construct invented by Europeans and used primarily to lend a scientific rationale to colonial policies, to a racism that was central to the British colonial empire. Scholars in the humanities and social sciences claim evidence of Aryan migrations—the emergence of the motif of the horse and the chariot—while linguists stress the Indo-European linguistic group that includes English and Sanskrit. Archaeological evidence of the presence of the horse in ancient remains continues to erupt at regular intervals (Daniyal 2018). How else could we explain these historical facts? For example, the historian Romila Thapar (1995: 95–96) argues that "the notion of the Aryans being a physical people of a distinct biological race, who moved *en masse* and imposed their language on others through conquest, has generally been discarded" in favor of smaller migrations. The Hindu nationalist movement in particular has used this alternate view of migrations with great enthusiasm. According to this argument, Indians are endogenous to South Asia, and there was never any significant Aryan migration into India and therefore no Aryan influence on South Asia. According to this version, the Aryan Migration Theory (AMT) is a colonial conspiracy that erases India's glorious precolonial history. Both versions circulate in the culture's public imagination (Malhotra 2011; Perur 2013; Chavda 2017; Joseph 2017). These competing stories are neither trivial nor innocent. They are deeply political, cultural, and critical threads that weave national identities—of where we come from, of who we are, and of where we want to go. The theories have profound consequences for questions of whether Indians are different, and if we are, what meaning we give that difference. This narration of history, the desire to belong, the recovery and reconstruction of a homeland are very evident in modern Hindu nationalism. Through the dual processes of the globalization of Hinduism by Hindu nationalists in India and the celebration and dissemination of Hinduism by diasporic Indians in their new lands lies the emergence of a global/universal Hinduism. This larger project is in many ways ultimately a propagation of *Hindutva* in the guise of Hinduism, expanding the quest for a Hindu nation through the help of Hindu diasporas across the globe. As Peter van der Veer (1994: xii) suggests, "Instead of encouraging a sense of world citizenship, the transnational experience seems to reinforce nationalist as well as religious identity."

DNA is the "master" molecule of our times, a molecule that has becoming increasingly central to the stories we tell about human bodies and their histories (Shea 2008). It has been called the "holy grail," the "book of life," the "blueprint of life," and the "secret of life" (Subramaniam 2013). DNA sequencing that retells origin stories has become a veritable cottage industry, dubbed "recreational genetics." Ancestry tests and genetic tests for disease susceptibility do a brisk business these days (Bolnick et al. 2007), even though the veracity, accuracy, and usefulness of these tests are the topic of numerous debates (Bolnick et al. 2007; A. Nelson 2016; Reardon 2017).

In particular here, I focus on the colonial roots of these stories. Spencer Wells, in his documentary *Journey of Man* (PBS Home Video 2002), encapsulates the central tension:

> What I'd like you to think about with the DNA stories we're telling is that they are that. They are DNA stories. It's our version as Europeans of how the world was populated, and where we all trace back to. . . . Our ancestors didn't pass down the stories. We've lost them, and we have to go out and find them. We use science, which is a European way of looking at the world to do that. You guys don't need that.

Starting with the failed human Genome Diversity Project, Jenny Reardon and Kim TallBear (2012) argue that DNA has been increasingly implicated in colonialist politics, rendering DNA a "Western" molecule, one that colonizes the lives and economies of native and third-world populations. Alongside, but in a politically unequal playing field, we have seen the rise of "people science," Do-It-Yourself (DIY) science, and "citizen science" movements, as well as the rise in movements that resist the colonizing impulses of science by insisting that DNA and its sovereignty be protected (Benjamin, 2009, 2013; Sur and Sur 2008; Reardon 2009; Subramaniam 2013). DNA has also emerged as a versatile molecule with "epistemic dexterity" (Benjamin 2015) and has been used as evidence for a wide variety of claims and ideologies. It would seem that it is "everyone's molecule," a powerhouse chemical lending its power to a breathtaking array of causes, ideologies, and politics. "People are trapped in history and history is trapped in them," James Baldwin wrote in 1955 in *Notes of a Native Son*. In the methods of DNA ancestry studies, our histories are trapped, quite "literally" and "materially," in the helices of DNA.

In exploring the politics of DNA and genomics in India, I examine, in three acts, claims of unique Indian biologies using the power of DNA. These three cases are purposely chosen from different sites and across different scales—local, national, and international—to highlight the wide scope of genetic nationalism. I explore the actual genetic studies and what they said, who deployed them and why, what claims about Indian biology they made, and the political stakes of these stories. These debates spilled into academic, popular, political, national, international, and activist circles because of the profound claims being made.

Act 1, A Global Brotherhood of Indians, is based on a 2001 genetic study that claimed that upper-caste Indians were more European than those in lower castes. Dalit activists used this evidence to argue that this study implied that caste was race, and therefore they should be allowed to take the case of caste discrimination to UN World Conference Against Racism and invite international monitoring. This case highlights marginalized groups' active uses of genetic evidence to subvert nationalist claims of a unique India.

Act 2, An Indigenous Nation Transnationally Bound, takes us across the ocean to the state of California. Here, I explore how a different set of genetic studies bolstered the claims of diasporic South Asian Hindus in California who challenged representations of Hindu and Hinduism in sixth grade textbooks. In this case, diasporic groups in part used genetic evidence to consolidate a unique India and Hindu nationalism.

Finally, act 3, Genomics *Swadeshi* Style, explores two recent developments in India. First up is the development of the Indian Genome Variation Project, a large genomic initiative that samples a diverse range of individuals and populations across the country to create a national genetic database. Another set of projects work to repackage indigenous medical systems in the language of genomics. The government has created a ministry of AYUSH to mainstream Ayurveda, Yoga and Naturopathy, Unani, Siddha, and Homeopathy with the "health care delivery system in India" (Shrivastava, Shrivastava, and Ramasamy 2015). Efforts such as Ayurgenomics attempt to reframe Ayurveda as a twenty-first-century modern medical system fully compatible with modern genetics. Both these projects are securing the future modernity of India through a bionationalistic embrace of genomics.

A short primer on caste: The biology of caste emerges as a central focus of these genetic studies. The caste system in India is as complex as it is contested (R. S. Singh 2000). The term *caste* is believed to have been introduced by the Portuguese in the sixteenth century (R. S. Singh 2000). It is an elaborate system with a long history. Overall, it has two elements: *jathis*, the endogamous community often defined by occupation and geography, and *varna*, a broader category of contemporary castes (Bamshad et al. 1996). The varnas outline four main castes: Brahmin, the priestly caste; Kshatriya, the warrior class; Vaisya, the merchant class, and Sudra, the lowest of the four classes. The caste system further splinters into subcastes. India is said to have about three thousand caste groups and tens of thousands of subcastes. The Dalits, often called the "untouchables," were considered outside the caste system, though in some schemes they are included as a fifth class, or Pancham (Mountain et al. 1995).

The caste system has been touted as the "largest social system ever designed," producing distinct inbreeding communities (R. S. Singh 2000). In practice, it is the defining factor that affects individual and groups in India throughout their lives. It touches all aspects of life—religious, social, economic, and political. The caste system evolved as populations grew, and with time it has become complex, intricate, and elaborate. It is rigidly defined and perpetuated by a division of labor and specialization of professions on the one hand and religious and commensal practices and marriage rules (endogamy) on the other (P. D. Gupta 2015). There are innumerable theories on the origins and evolution of caste (Ghurye 1969; Kosambi 1988; Dirks 2001; Desai and Dubey 2012; S. Guha 2013), ranging from ones that stress its ritual-religious aspects to ones that stress its political economic aspects. Increasingly, social scientists have traced caste's shifting political "avatars" (Srinivas 2000; S. Guha 2013). It is striking that despite this scholarship, the scientific impulse to locate caste as a biological characteristic persists. This chapter explores the contours of the biologization of caste.

ACT 1. A GLOBAL BROTHERHOOD OF INDIANS

The story (as I narrate it) begins in India with a June 2001 headline in *Frontline* ("India's National Magazine") proclaiming, "New genetic evidence for the origin of castes indicates that the upper castes are more European than Asian"—a resurgence of a centuries-old debate on caste and race (Ramachandran 2001). Was caste really race in a different guise in the social fabric of India? The current controversy, in keeping with the long history of scientific studies on race, hinged on a study that purported to link race and caste. An international group of scientists led by Michael Bamshad of the University of Utah's Eccles Institute of Human Genetics, along with scientists from Estonia, India, the UK, and elsewhere in the United States, had just a few months earlier reported finding evidence that India's upper castes were more closely related to Europeans and lower castes more closely related to west Asians (Bamshad et al. 2001). As one of the authors, Lynne Jorde, put it, "Groups of males with European affinities were largely responsible for this invasion 3,000–4,000 years ago" (quoted in Cooke 1999). Interestingly, Dalit groups used this "scientific" evidence to argue that contrary to traditional claims, caste was analogous to race and therefore merited a hearing at the United Nations World Conference Against Racism, to be held in Durban in late August and early September 2001. This study has been well examined (Viswanathan 2001;

A. Pinto 2001b; Sabir 2003; Reddy 2005; Sur and Sur 2008; Loomba 2009; Subramaniam 2013). I rehearse the arguments here to highlight the diverse contexts of genomic bionationalism.

In their study, Bamshad and his colleagues sampled 265 males from eight different castes and ranks from one state in the south of India, Andhra Pradesh. They compared several markers in the mitochondria, Y chromosome, and autosomal DNA to extant databases of 400 Indians and 350 Africans, Asians, and Europeans. The scientists highlighted three main results. First, for maternally inherited mtDNA (mitochondrial DNA), each caste is most similar to other Asians. However, 20–30 percent of Indian mtDNA haplotypes belong to West Eurasian haplogroups, and the frequency of these haplotypes is proportional to caste rank, with the highest frequency found in the upper castes. Second, paternally inherited Y chromosome variation in each caste is more similar to Europeans than to Asians. Moreover, in autosomal DNA the affinity to Europeans is proportionate to caste rank, with the upper castes most similar to Europeans. Finally, because of the differences between maternal and paternal inheritance patterns, they argue that these results suggest an upward mobility in the caste system that women, but not men, could achieve.[3]

At the heart of this portrayal is a story of human migration. Who are Indians? Where did they come from? Dalits, members of an oppressed group of India's caste system, wanted to take the case of the system's continuing brutality to the United Nations World Conference Against Racism. They argued that caste discrimination against Dalits (a term that literally means "crushed and broken") has continued to the present day, affecting over 160 million in India. The Indian government must be held accountable, and the matter needed international attention and monitoring. According to Smita Narula (2001) of the international group Human Rights Watch, "caste discrimination was a 'hidden apartheid' and affected more than 250 million people in India, Nepal, Sri Lanka, Bangladesh, Pakistan and Japan." The Indian government disagreed. They argued that caste was not race and therefore ought not to be discussed at the UN conference on racism. Indian leaders have long declared caste to be an internal and "national" issue, not one that international agencies ought to monitor. Hindu nationalist leaders closely associated with the BJP added fuel to the fire by "declaring that caste was a part of India's ancient traditions—and could not be discussed at international fora" (Devrag 2001).

While the Dalits fought to include caste discrimination in the conference and the government fought against its inclusion, intellectuals joined in on both sides of the debate (Omvedt 2001a, 2001b). Some argued the government's position, others the Dalits', and still others argued that while caste discrimination was a horrendous practice, we must be intellectually honest and not confuse "race"—already a biologically dubious concept—by equating caste with race. Social anthropologist Andre Beteille resigned from the National Committee on the World Conference Against Racism in protest when the committee began drafting a presentation to the conference. Beteille (2001) suggested that treating caste as a form of race was "politically mischievous and scientifically nonsensical." He concluded, "We cannot throw out the concept of race by the front door when it is being misused for asserting social superiority and bring it again through the back door to misuse it in the cause of the oppressed" (Beteille 2001).

The strategy of claiming links between race and caste have long been embraced by Dalit activism and resistance, as the Dalits have redefined themselves as "the black untouchables of India" and built solidarity with black resistance movements across the world (Rajshekar 2009). To understand the politically charged nature of caste in these debates, we need to understand the increasing centrality of caste politics in India. The rise of Dalit and other "lower" caste–based parties is a phenomenon to be reckoned with in contemporary India (Varshney 2000; Jaffrelot 2003). Dalit activists have tirelessly campaigned against deep-rooted and widespread discrimination they continue to face. Given the entrenched politics of caste in India, Dalit activism has long sought international alliances, and activists have continued to internationalize their struggles. The Dalit Panthers emerged as a militant organization of Dalit activists in Maharashtra in the 1970s. They borrowed their name, with its insurrectionist symbolism, from the Black Panthers of the United States (Omvedt 1995). The name also invited association with a global imaginary of progressive literature and politics—the Black Panthers, anti–Vietnam War protests, the 1968 Paris protests—as part of a larger antiestablishment aesthetic (Rao 2009). Their 1973 manifesto makes clear that they expanded their group to include other oppressed minorities such as landless laborers, poor peasant women, and scheduled castes and tribes (Azam 2017). Drawing on long histories of injury, they have adopted a radical politics and a symbolic appropriation of violence (Jaoul 2013). Anupama Rao's *The Caste Question* presents an incisive analysis of Dalit politics in modern India, arguing that Dalits reentered the realm of public space and self-presentation by adopting a "politics of presence" with the Dalit Panthers. Atrocities against Dalits played a role in the organization's advocacy of counterviolence. The Dalit Panthers often responded to news by rushing to the scene with cycle chains, knives, and wooden staffs, appearing ready for a street fight if necessary. Indeed, intense street fighting has occurred as Dalits have asserted their rightful role and place in the Indian polity (Rao 2009).

In internationalizing Dalit politics, the genetic studies that brought together categories of caste and race offered an opportunity for political alliances between anticasteist and antiracist activism. In using the Bamshad et al. study, the Dalit groups alleged that the evidence that caste was linked to race supported the century-old link between "black" activists across the globe who alleged a global racism against blacks. There was no indigenous and unique "India," this genetic data suggested, but rather a global brotherhood and sisterhood of caste and racial structures. Indeed, Dalit activists argued that the genetic evidence supported a global black brotherhood—linked not only by skin color but also now through shared genetic kinship, and through the long-enduring oppression of race and racism.

The Indian state, for its own part, worked hard in this case to retain control over its sovereign interests and governance (Omvedt 2001b). They predictably worked to make this a domestic and social issue rather than acknowledge its transnational dimensions. Finally, the Hindu nationalists have long claimed India as an ancient, unique, and glorious Vedic civilization and India as the cradle of humanity. Their embrace of genetic studies that purport the opposite of the Bamshad study—an indigenous and exclusive Indian origin—supports centuries of such activism. Thus, different genetic studies predictably found reception among different political groups, furthering their long-standing claims.

Scientists, especially in migration studies, have embraced genetic technologies with renewed gusto. It is important to stress that many Indian geneticists not connected with Hindu nationalism have also embraced genetic technologies as productive sites to empirically adjudicate long-disputed social issues (Thangaraj, Ramana, and Singh 1999; Basu et al. 2003; Reddy et al. 2007; Majumder 2010; Majumder and Basu 2015; Tamang, Singh, and Thangaraj 2012; Analabha Basu 2016). Large global genetic data projects are funded by private companies and by governments. Genetic testing and access to genetic information—whether for migration studies, population sampling, or personal genomics and health—have been welcomed, as DNA sequences have been embraced and implicated in projects with a breathtaking array of projects and goals (A. Nelson 2016; Reardon 2017).

On the other hand, many theorists and social activists, including Dalit activists, point to both race and caste as potent systems of social stratification. As Ambrose Pinto (2001a) remarks, "Prejudice and discrimination are both a part of caste and race. And what is worse is that such prejudice and discrimination are not merely personal but institutional, a part of the structure and process of the whole society. In both caste and race theories, the so-called higher or superior groups take the attitude that their culture is superior to all other cultures, and that all the other groups should be judged according to their culture. What is the difference between the claims made by the white race in Europe and the upper castes in India?" Similarly, while scholars continue to claim race as a "global" construct that categorizes people across continents (a claim repeatedly debunked), claims of caste remain localized within the Indian subcontinent. An illustrative example Dipankar Gupta (2000) cites is endogamous marriage, which always takes place within castes of the same local communities but not between members of the same caste across regions, since these are not considered analogous. He also stresses the many differences between race and caste, such as the ways in which the idea of pollution—a central element of the caste system—is largely absent from race.

Scholars have pointed out that the academic debates around the UN World Conference Against Racism were largely a discussion among upper-caste Indian academics who had long sought to Indianize caste, keeping it as a unique Indian attribute rather than globalizing it as one of many kindred structural systems of oppression that spanned continents and nations (Viswesaran 2010). The academic discussions reveal the fraught politics of caste in India. After months of public debate and deliberations, the Indian state won. Caste was not included as a topic at the UN conference.

As we shall see, genetic analyses on populations have proliferated in the decades following Bamshad et al. Conflicting patterns surface in the various studies, lending themselves to multiple interpretations. Genetic technologies have thus become an important site of identity debates, as various groups have embraced particular genetic data sets to bolster their political claims (Heath, Rapp, and Taussig 2004; Rose and Novas 2004).

ACT 2. AN INDIGENOUS NATION TRANSNATIONALLY BOUND

Half a decade later and an ocean away, two groups of diasporic South Asian Hindu activists in the United States, the Vedic Foundation and the Hindu Educational Foundation, challenged the representation of Hinduism in sixth-grade textbooks in California in 2005–2006 (Bose 2008; Viswesaran et al. 2009). Challenging the content of textbooks has been a powerful tool in social movements across the globe. Controlling what children learn solidifies knowledge claims and shapes knowledge structures in the years to come.

The representation of Hindu and Hinduism in school curricula and textbooks in particular has been a controversial and contested site of activism. When Hindu nationalists came to power at the national level in India in 1998, they established the National Curriculum Framework to reshape textbooks. Earlier frameworks that were grounded in democratic values, social justice, and national integration were reshaped to offer a "value education" that inculcated a "national spirit" and generated pride about India's heritage (Viswesaran et al. 2009). In India, Hindu nationalist–headed state governments have undertaken extensive curricular and textbook reform in the guise of improving educational standards (Hasan 2002; Bose 2008). In the United States, the representation of Hinduism in textbooks also engages with multicultural politics of inclusion, making arguments of Hindu nationalism versus Hindu inclusion difficult and slippery territory (Kurien 2015). Within US multicultural politics, Hindu activists argued that Hindu American children "suffer low self-esteem and confidence when faced with the histories of caste and gender discrimination in South Asia" (Sengupta 2016).

Among the many representations the California Hindu activists challenged were the portrayals of Hindu gods and mentions of the subordination of women in Hinduism and the oppressive nature of the caste system. At the heart of Hindu nationalist claims is the consistent conflation of India with South Asia, and Hindu India with a multireligious India. The claims of a unique indigenous India include a revival of a long-enduring debate on the Aryan Migration Theory (AMT). Was there an Aryan migration, or is this an orientalist mythology? The California School Board was asked to adjudicate these claims. While a small part of the larger debate, it is interesting that the claims included genetic data. In support of removing the AMT from textbooks, they cited a pair of genetic studies that contradicted the Bamshad study—a 1999 study by Kivisild et al. and a 2006 study by Sahoo et al.—to claim that the Aryan Migration Theory was "conclusively disproved."

In the Kivisild study, the authors focused on language groups as populations, sampling 86 Lambadi, 62 Lobana, 12 Tharu and 18 Buksa, and 122 predominantly Indo-Aryan language speakers from Uttar Pradesh, along with 250 Telugu (Dravidian language speakers), and used these language populations to compare 550 polymorphic genetic markers on mitochondrial DNA. The study concluded that there was "deep" and common ancestry of Indian and western-Eurasian mtDNA, suggesting that the "deep ancestry" could be traced to the Pleistocene rather than more recent migrations claimed by the Aryan Migration Theory. Sahoo et al. (2006) similarly conclude:

> The Y-chromosomal data consistently suggest a largely South Asian
> origin for Indian caste communities and therefore argue against any
> major influx, from regions north and west of India, of people associ-
> ated either with the development of agriculture or the spread of the
> Indo-Aryan language family. The dyadic Y-chromosome composition
> of Tibeto-Burman speakers of India, however, can be attributed to
> a recent demographic process, which appears to have absorbed and
> overlain populations who previously spoke Austro-Asiatic languages.

These studies suggested that there was no recent outside migration into India and posited an indigenous and exclusive genetic citizenship for all contemporary Indians (Sahoo et al. 2006). They argued that genetic evidence superseded linguistic, anthropological, and historical claims of an Aryan migration. Likening this evidence to biblical philological evidence, a guest columnist at Stanford University argued that ignoring this scientific evidence was tantamount to sanctioning creationism! Secular academics and activists, on the other hand, challenged the evidence on many grounds. The California Board of Education was asked to rule. Were Indians a "pure" population? Was there no Aryan migration? Ultimately, after numerous hearings and the involvement of academics from all over the country, some changes were adopted, but the Aryan Migration Theory remained in sixth-grade textbooks (Burress 2006).

In 2016 a new claim emerged, and activists worked to address many issues they had raised before that were not taken up by the Board of Education. Again, the Aryan Migration Theory emerged as a point of contention. They cited two new articles, Vemsani (2014) and Perur (2013), as "proof" that genetic studies have debunked the Aryan Migration Theory. However, a closer analysis of these articles and other, newer studies show that this is far from true.[4] None of the authors the articles cited actually claim to have solved the question of whether there was an Aryan migration (Joseph 2017). Vemsani (2014) repeats earlier arguments that since there exists "scant anthropological and linguistic evidence for Aryan migration, genetic studies can be used as scientific evidence." However, most genetic studies are limited by small sample sizes and a small subset of available genetic markers, as well as by assumptions they have to make for statistical analyses. For example, many analyses assume that migrations involve genetic mixing of homogeneous populations when in fact founder groups are rarely homogeneous. In this sense, genetic evidence cannot "step up" to explain events where archaeological and linguistic evidence may be weak or nonexistent. The second article they cited, by Perur (2013), is based on an interpretation of a study by Reich et al. (2009) that used samples from 132 Indians across 25 diverse groups to estimate the "founder groups" of current Indian populations. They concluded that there were two main founder groups, which they named Ancient North

Indian (ANI) and Ancient South Indian (ASI), and that the degree of mixture varied between 40 and 70 percent. Further, higher ANI markers were associated with higher caste groups, and ANI was linked to markers also present among current-day Europeans, while ASI was more closely linked to the Onge (Andamans). These results supported a study by Moorjani et al. (2013) that concluded that the ANI-ASI mixture dates ranged from 1,900 to 4,200 years ago. Overall, neither study proved or disproved any migration theory—except for suggesting the possibility of an influx of a new population (ANI). But as even the authors note, the studies had access to small sample sizes; more comprehensive studies, with extensive sampling of Indian across regions, would be necessary for more robust conclusions.

Importantly, genetic studies largely substantiate theories already in existence. As geneticist Aravinda Chakravarti remarked, "To a cynic, the existence of the ANI or ASI, their unique and remote ancestry within India, or their suggestive identities as Indo-European and Dravidian speakers, are already common knowledge" (A. Chakravarti 2009). After a hearing, many changes were adopted, but the caste system and the Aryan Migration Theory remained in the textbooks (Medina 2016). In 2017, a group calling itself California Parents for the Equalization of Educational Materials sued the California Board of Education, alleging that the state's history standards "negatively portray Hinduism" (Yap 2017). Given that Indian Americans are one of the most highly educated and wealthiest groups in the United States, and that 79 percent report that religion is important to them, no doubt such challenges will continue (Andersen and Damle 2018).

Following this challenge, David Reich and a group of 92 scientists authored a new study in 2018, "The Genomic Formation of South and Central Asia," which found evidence supporting the Aryan Migration Theory (Narasimhan et al. 2018). *The Economist* (2018) proclaimed, "A new study squelches a treasured theory about Indians' origins" and concluded from the paper that "the Aryans did not come from India, they conquered it." But undoubtedly the debate will persist as studies continue to explore the genomes of the world. As Ashok and Roychowdhury (2018) asked of this recent study, "Will it settle or again trigger the contentious debate?" But what is clear in both these instances is that a mobilized Hindu diasporic community is actively working to carefully craft the representations of Hinduism. We are likely to see many more such challenges.

While in acts 1 and 2 I have highlighted the particular genetic studies that were used in the two cases, genetic migration studies of India are virtually an industry, and the genetics of Aryan Migration continues to be hotly debated as new studies emerge with new results (Joseph 2017; Danino 2017). What is clear in the rise of such studies is that the debate on origin stories has moved to the genomic level, as the "scientific" mode of contemporary India. Scholars across the political spectrum embrace genomic studies, even as they issue cautions about their reach and use. Through genomics they seek to resolve centuries-old debates and build a new genomic infrastructure of the peoples of India. Such work bolsters the epistemic authority of genetic/scientific evidence as superior to methods and results from the humanities and social sciences. Predictably, activists preferentially embrace data sets that support their position while ignoring others. Finally, and in keeping with a long history of the political use of DNA evidence (Nelkin and Lindee 2004), there is often little nuance in recognizing the limitations of genetic studies. Most studies inevitably have to work with whatever markers and data sets are available, and they always must include many assumptions for their statistical models. As the genetic study by Reich et al. (2009) warns us:

> "models" in population genetics should be treated with caution. Although they provide an important framework for testing historical hypotheses, they are oversimplifications. For example, the true ancestral populations of India were probably not homogeneous as we assume in our model, but instead were probably formed by clusters of related groups that mixed at different times.

It is also important to stress that migration studies is a thriving scientific field. Many secular scientists from India with no open ideological commitments to nationalism have ardently embraced these methods as a way to shed light on the peopling of India (Thangaraj, Ramana, and Singh 1999; Basu et al. 2003; Reddy et al. 2007; Reich et al. 2009; Narasimhan et al. 2018; Tamang, Singh, and Thangaraj 2012; Majumder and Basu 2015; Analabha Basu 2016). What gets elided repeatedly in the political sphere is the need for interdisciplinary work. Debates routinely assume that there is only one correct history and that we can access it. We need interdisciplinary analyses to explore the complexities of human migrations using all data and lines of evidence (genetic, linguistic, anthropological, sociological, philosophical, archeological, and historical) available to us. After all, genetic histories may not correspond with linguistic or archaeological histories. Languages may diffuse differently and leave different marks on history than genes do (Creanza et al. 2015). Languages are inherited "vertically" (from parents to children) like genes, but they are open to more vigorous transformation horizontally and laterally based on contact among populations. Gendered histories that rely on maternal and paternal lines are fraught methodologically (Morris and Lightowlers 2000; Margulis and Sagan 2003). As long as they are methodologically rigorous, might not each of these partial histories give us useful information? Yet in our contemporary times, the "truth" of history has been ceded to the power of DNA and genomic analyses. Given the repeated historical contestations of the Aryan Migration Theory for several centuries now, genomic nationalism is unlikely to be the last word on the ongoing debates of India's prehistory.

ACT 3. GENOMICS *SWADESHI* STYLE

In recent years, India has promoted a unique *swadeshi* or indigenous "made in India," Indian "body" or biology, owned, patented, and mined by India for Indians. The slogan that has enlivened this project is "Unity in Diversity." India is a diverse country with a multitude of variation, the claim goes, yet there is unity within its borders. Thus, bionationalism is able to incorporate both the nation-state and its borders and claims of difference and variation within those borders (Oikkonen 2014). Yet in creating a "representative" database of India's genome, the project relies on familiar modes of difference in its sampling techniques—region, religion, caste, subcaste, community, and so on—and as a result, once again, biology inevitably reifies, reinforces, and reproduces old categories of difference as valid biological categories of variation for modern India. The techniques are also distinctly gendered, told through the gendered geographies of the Y and X chromosomes (Nash 2004, 2012).

Several sets of global genomic developments shape the scope and structure of the project. The Human Genome Diversity Project, proposed in the early 1990s, soon encountered numerous problems. It was seen as an imperialist and colonizing project that sought to "steal" DNA from native populations across the globe with improper consent (Reardon 2009), and it was dubbed the "Vampire Project" in some circles. The profound consequences and political importance of DNA have exploded in the last two decades. As a result, many countries of the Global South and native populations across the globe are wary of submitting to international genomic initiatives that seek to collect DNA. In these techno-scientifically mediated times, there are concerns about privacy and fears about how information may be saved and used. Second, many emerging global powers outside of the United States and Europe have moved to control the security of their nation's or continent's DNA by taking control: Mexico (Silva-Zolezzi et al. 2009), Africa (Gurdasani et al. 2015), and several Asian countries, including Singapore (Teo et al. 2009), Taiwan (Fan, Lin, and Lee 2008), and Korea (Ahn et al. 2009). The consolidation of nations through claims of shared genomes (Underhill et al. 2010; Haber et al. 2012) bolsters genomic nationalism. India has also entered modern genomics with massive state-funded projects. These projects have strong backing from their national governments, creating new linkages between genetic identities and national sovereignty (Subramaniam 2013). The emerging biotechnology industry in Asia has embraced bionationalism and is quickly reshaping the global development of genomics, as well as insisting on their own "genomic sovereignty" (Benjamin 2009; Kelly and Nichter 2012; Ong and Chen 2010). After all, the right to control or own one's biological constituents is at the heart of deeply ingrained liberal commitments to self-governance (Reardon 2017). By asserting genomic sovereignty, such bionationalism that seeks to secure and control their national genomics have frustrated more global attempts. India joins this trend.

I want to focus on two projects that capture the growing focus of the Indian state. First, I explore the creation and support of the Indian Genome Variation Project, which parallels the efforts of other national and international genome variation projects. Second, I examine the attempt to geneticize indigenous medical systems like Ayurveda through projects such as Ayurgenomics.

The Indian Genome Variation Project

The Indian Genome Variation (IGV) Initiative, established in 2003, involved six constituent laboratories of the Council for Scientific and Industrial Research and funding from the Government of India. The members of the consortium are the Institute of Genomics and Integrative Biology, Delhi; the Centre for Cellular and Molecular Biology, Hyderabad; the Indian Institute of Chemical Biology, Kolkata; the Central Drug Research Institute, Lucknow; the Industrial Toxicological Research Centre, Lucknow; and the Institute of Microbial Technology, Chandigarh.

This ambitious project is the "first large-scale comprehensive study of the structure of the Indian population" (Narang et al. 2010) with wide-reaching implications. As the project argues, India is a large, populous, and diverse on many levels. It comprises of "more than a billion people, consists of 4693 communities with several thousands of endogamous groups, 325 functioning languages and 25 scripts." The project argues that to "address the questions related to ethnic diversity, migrations, founder populations, predisposition to complex disorders or pharmacogenomics, one needs to understand the diversity and relatedness at the genetic level in such a diverse population" (IGVC 2005). The project has been touted as one of disease gene exploration (IGVC 2008). They have identified over a thousand genes to study, "selected on the basis of their relevance as functional and positional candidates in many common diseases including genes relevant to pharmacogenomics" (IGVC 2005).

The collection of genetic samples to map the demographics of health, however, has found resistance from nations across the globe. HapMap projects have sought an inventive route around this. They have sought to get DNA through diasporic communities, treating such communities as Diasporic Proxies (Hamilton 2015). For example, as Hamilton (2015) explains, in not having access to Indian DNA, HapMap 3, the third phase of the international HapMap project, included a group of Gujaratis living in Houston as a sample population. Diasporic Indians of Gujarati descent therefore come to stand in for "India." Given the woefully poor sampling methods in representing whole nations in the larger project, this is not unusual. The Indian Genome Variation Project's framing and scope undercut and challenge the usefulness of such diasporic proxies, since the IGVP is predicated on India's population as regionally "diverse" and substructured.

The Indian project joins a global shift in turning human health into a biotechnological project, with a specific end goal, a pharmaceutical solution, ushering in the "pharmaceuticalization" of life. International genomic efforts such as the HapMap projects are interested in the global distribution of genomic variation. The development of biotechnology—infrastructure, methods, instruments, scientists, methods, and data—has become a critical site of capital. With the onset of such investments, India has arrived as an international player and an emerging power in biopolitical governance. In particular, with the geneticization of biomedicine, the global distribution of health problems is translated into and reduced to genetic susceptibility maps. These aspirations are very much linked to the pharmaceuticalization of medicine (Pollock 2014) and the development of pharmacogenomics, whereby genetic susceptibilities spawn new classes of drugs. Countries such as India and Mexico are seen as "Pharma's Promised Lands" (Benjamin 2009). As Sunder Rajan (2017) argues, we live in an era of "pharmocracy." Pharmocracy is the culmination of a longer trajectory of opening up markets, global patent restructuring, and the pharmaceuticalization of medicine. In India, as Sunder Rajan (2017) demonstrates, this emerged through a process of "harmonization." It began in 2005 when Indian pharma regulations came into global compliance with intellectual property and clinical trials, and culminated in the current moment with an "expansion of multinational corporate hegemony."

These recent developments are the newest phase of a larger historical trend toward the "molecularization" of life (Rabinow and Rose 2006; Rose 2006; Egorova 2013). Organisms' biologies, rather than being considered at the organismal level and in the context of their environments, are increasingly reduced to their molecular selves. The Human Genome Project, the HapMap Project, 1000 Genomes, National Geographic's Geographic Projects, and consumerist companies like 23andMe, Ancestry.com, and Veritas Genetics all molecularize life. As critics reveal, this has shifted our conceptions of ill health and disease from a focus on the social contexts of poverty and access to nutrition and care to a focus on genetic propensities to ill health or disease (J. Kahn 2013; Dumit 2012; Chambers et al. 2014).

In its embrace of "biocapital" through bionationalism, the IGV project hopes to compete in the global and international marketplace. In mapping India's genetic diversity, the initiative picks global standards and markers. In the Indian context, these emerge from colonial-era medicine and census categories that demarcated population data along the lines of ethnic groups, caste groups (and subcastes), state and regional categories, and language groups. Given the analyses of such projects in the West, where old politics of race, gender, and nation reappear in modern genetic language (Pálsson 2007; Bell and Figert 2015; Benjamin 2015; Hamilton, Subramaniam, and Willey 2017), there is no reason to believe that India will yield anything different. A plethora of studies in national and transnational collaborations have begun chronicling the susceptibilities of population groups across India to diseases such as diabetes, cancer, hypertension, heart disease, Alzheimer's, and Parkinson's disease (Jamal et al. 2017; Kumudini et al. 2014; Saxena et al. 2013; Singh et al. 2014; Tan et al. 2014; Venkatachalam et al. 2015).

With a backdrop of colonialism and the continued center of power in the "West," India is asserting its biopolitical independence, nurturing local talent and building a strong national and transnational infrastructure. Yet it is striking that this investment is happening in a country that lacks very basic health infrastructure, a nation still reeling from extreme poverty, where preventable and communicable diseases are responsible for most of its citizen deaths (Najar and Raj 2015). India's health statistics are abysmal even after decades of strong economic growth; less than 1 percent of the nation's GDP is spent on public health care, and it has only nine hospital beds per ten thousand people (in comparison, China has a rate of forty-one beds per ten thousand people) (*New York Times* 2014).

The Indian government has sunk millions of dollars into sequencing the Indian genome. The rhetoric of the "Indian Genome Project" is predicated on DNA technology that aids the health and well-being of its citizens through sequencing genomes and uncovering disease vulnerabilities, where genetic landscapes of India provide "a canvas for disease gene exploration." It is clear that the logic of the IGV initiative is constructed around health and drug delivery, that is, translating the diverse body types of India into a plan for delivering individualized medical solutions (Hardy et al. 2008). The logic is well articulated in this abstract:

India currently has the world's second-largest population along with a fast-growing economy and significant economic disparity. It also continues to experience a high rate of infectious disease and increasingly higher rates of chronic diseases. However, India cannot afford to import expensive technologies and therapeutics nor can it, as an emerging economy, emulate the health-delivery systems of the developed world. Instead, to address these challenges it is looking to biotechnology-based innovation in the field of genomics. The Indian Genome Variation (IGV) consortium, a government-funded collaborative network among seven local institutions, is a reflection of these efforts. The IGV has recently developed the first large-scale database of genomic diversity in the Indian population that will facilitate research on disease predisposition, adverse drug reactions and population migration. (Hardy et al. 2008: S9)

The article also outlines the strategy for such a project. The authors clearly tout the goal of making India a more attractive and available site for global capital exploitation. As Sunder Rajan (2006) explains, biocapital is increasingly shaped by the marriage of biotechnology and market forces. The IGV project explicitly positions India as a hub, poised to present its population as a resource for medical experimentation:

India has thus positioned itself as a global hub for conducting clinical trial research by investing in capacity and infrastructure. Some domestic companies subsidize their research and development (R&D) platforms by providing contract services for clinical trials to multinational and foreign companies. Accordingly, a "predictive population database" could help maintain India's competitive edge by improving the selection specificity through stratification of the test population, thereby further reducing the time and cost associated with conducting clinical trials in India. (Hardy et al. 2008: S10)

The Indian Genome Project argues that DNA technology will aid the health and well-being of its citizens through sequencing genomes and uncovering disease vulnerabilities. Feminist scholars of medicine have amply demonstrated the health industry's recent shift toward transforming health care into a product, with tests, drugs, and predictive technologies for the healthy (Dickenson 2013; Swanson 2014; Foster 2017). One of the enduring lessons of the debates over biological determinism versus social construction is that biological determinism renders bodily conditions as "immutable" (Hubbard 1990). As a Foucauldian analysis of biopolitics would suggest, when ill health is a problem due to individual's genetic propensity rather than the inequities of an everyday life of poverty, polluted environments, and the lack of access to good health care, then the problem and the solutions shift from the state and public policy to the individual. As a result, pharmaceutical drugs are presented as a solution, rather than state investment in better infrastructures for improved access to good air, water, and nutrition. Strong political, economic, and ideological assumptions undergird these biotechnological assumptions and priorities. India joins an international wave of such shifts in medicine. In the United States, racial inequalities have been similarly rebiologized in recent times (Stepan 1982; Gannett 2001; J. Kahn 2005; Reardon 2009; Hammonds and Herzig 2009). Despite projects like Swachh Bharat (Clean India), inaugurated with much fanfare to clean India's environment and increase accessibility to cleaner air and water, the results have been disappointing (Sagar 2017). For example, unprecedented air pollution has continued to make international news. In November 2017, the pollution levels in some cities reached "levels nearly 30 times" what the World Health Organization deems safe, creating a "public health emergency" (*New York Times* 2017). It is estimated that about 2.5 million people in India were killed in 2015 because of the pollution levels, more than in any other country (Das and Horton 2017).

The gendered dimensions of medical technological development are striking. In a country where women feed themselves last, women's health statistics are particularly horrendous. More than 90 percent of adolescent girls in India are anemic, and 40 percent of Indian mothers are underweight (G. Harris 2015). Global public health organizations present India's poor as one of the most abject populations in the world. Despite evidence that genetics plays little role in India's demography of ill health (G. Harris 2015), the funding of a mega genomic project rather than one of public health infrastructure is remarkable (Subramaniam 2015a).

These shifts in biomedicine have imported biotechnological trends of the West, what Donna Dickenson (2013) calls "Me Medicine": practices that focus on an individual's needs and interests. Her analysis shows that rather than scientific plausibility, what has driven recent biomedicine and the diffusion and availability of products and services is the development of new markets, products, and services catering to individual needs and perceived threats and risks. She calls instead for a return to a "We Medicine" approach that emphasizes investment in public health infrastructure, which has already extended human life spans radically.[5] She argues that "We Medicine" that emphasizes technology used for the common good, coupled with better regulation of biotechnology industry, should be our path forward, restoring the idea of the commons in modern biotechnology (Dickenson 2013).

Geneticizing Indigeneity: Ayurgenomics

India has several indigenous systems of medicine, each with "specific purpose, logic and ethics" (Sujatha and Abraham 2012). There has always been a tension between indigenous or traditional systems of medicine and modern allopathic "Western" systems (often referred to colloquially as "modern medicine"). The former are routinely deemed religious and subjective and not accorded the validity of allopathy, and binaries persist—biomedical vs. ethnomedical, illness vs. disease, and epistemic knowing vs. gnostic knowing (Sujatha 2007). Despite often being labeled as a pseudoscience, Indian systems of medicine have thrived through "revivalism, syncretism and hybridization" to retain a pluralistic landscape of medical systems in India (Sujatha 2007).

In 2014 Narendra Modi, now the prime minister, ran on a platform that stressed Hindu nationalism and development, and he championed India's indigenous medical systems (IMS). Once in power, he elevated IMS as a separate AYUSH ministry (Ayurveda, Yoga and Naturopathy, Unani, Siddha, and Homeopathy), including a new minister of state charged with promoting "educational standards, quality control, and standardization of drugs" (S. Kumar 2014). In the first three years, sixty-four AYUSH hospitals have emerged across India, with the end goal of one hospital in each district in India (*Hindu* 2017). The prime minster also inaugurated the first ever All India Institute of Ayurveda in Delhi with the goal of bringing "synergy between the traditional wisdom of Ayurveda and modern diag-nostic tools and technology" (*Hindustan Times* 2017). He has called for private companies to invest in Ayurveda and find "medicines which can, like allopathy, give immediate relief to people" (*Times of India* 2017). In December 2017 the government introduced a National Medical Commis-sion bill, which overhauls the medical system, its governing body, medical education, and licentiate exams. If passed, it will repeal the Indian Medi-cal Council Act of 1956.[6] A controversial feature of the bill is a provision that would allow practitioners of Ayurveda and other traditional Indian systems of medicine to prescribe allopathic drugs after passing a "bridge course" (*Times of India* 2018). Indian doctors across the country went on strike to protest the new bill (*Financial Express* 2018).

With the growth of herbal medicine in the global health market start-ing in the 1980s, there is increased pressure to diversify and commodify IMS. Today IMS serve many interests, including nationalist, private, cor-porate, tourist, global, and local (Sujatha 2007; Mukunth 2015). Even multi-nationals have created Ayurvedic consumer products. However, much of the research is driven by the biotechnology sector, as pharmaceutical companies and international agencies and governments mine IMS for effective remedies. Scholars argue that there is little or no innovation in IMS or shifts in epistemologies of knowledge, only a push toward market-ing drugs (Sujatha 2011). The marketing of drugs is part of a global counter-culture marked by appeals to neo-orientalism and portrayals of Ayurvedic drugs as facilitators of spirituality (Bode 2015). The proliferation of these new medicines has unfortunately made them more expensive and inac-cessible for a population that is largely poor. Many of the new innovations have focused on fast-selling "cosmetic" products that target a gendered market, appealing to traditional forms of masculinity and femininity (Islam and Pearce 2013). Drugs reinforce men's sexual prowess and the "power of masculinity" (with drugs for premature ejaculation and erectile dys-function) while women's health is defined around beauty products. Thus, women's health is associated with nature and men's health with power.

Perhaps the most developed system in IMS is Ayurveda, an ancient system of medicine documented and practiced since 1500 BC (Prasher, Gibson, and Mukerji 2016). Indigenous systems, especially Ayurveda, have long been closely linked to nationalism in India (Alter 2015). Starting in colonial times, IMS were used to challenge colonialism. In challenging the colonial state, Ayurveda was absorbed into the inherently modern late colonial biopolitics and incorporated into the infrastructure of population management and strategies for development. Modern Ayurveda has been centrally incorporated into the project of population management (Berger 2013a, 2013b).

In Ayurveda, the basic constitution of a body, or *prakriti*, is constructed through three modules, or *trisutra* (etiology, signs and symptoms, and treatment or management of diseases). These *trisutra* are in turn determined by the principles of *tridosha*, three physiological energies: the *vata* (kinetic), *pitta* (metabolic), and *kapha* (potential). Ayurveda believes that these energies work in concert with the external environment to maintain bodily homeostasis (Prasher et al. 2017). Rather than highlight the conceptual and epistemological differences with allopathy, recent trends have worked to integrate genomic insights into Ayurveda toward visions of individually driven, personalized medicine (Aggarwal et al. 2010; Prasher et al. 2017). Ayurgenomics has worked to integrate the *trisutra* concept with genomics so as "to observe biochemical and molecular correlates of *prakriti*" (Prasher, Gibson, and Mukerji 2016), thus augmenting Western medicine (P. D. Gupta 2015).

Ayurgenomics exemplifies the vision of an archaic modernity. While the new government introduced a ministry for AYUSH and therefore the numerous medical systems of India, it is clear that some, like Yoga and Ayurveda, are particularly favored. The explosion of Ayurvedic medicine, personal products, and consumer goods shapes and is shaped by the growth of bionationalism and "biomoral consumerism" (Khalikova 2017).

Ayurveda invokes and provokes nationalism through various political and institutional practices that work though a politics of "cultural heritage," drawing on the histories of neo-orientalism and fundamentalism (Alter 2015). On the one hand, practitioners tout the indigenous roots of Ayurveda, highlighting its uniqueness—the concept of *prakriti* and its environmental, contextual view of the body. Yet the entire discourse around *prakriti*s, important to modern-day Ayurveda, were fairly marginal in classical Ayurveda and came to be magnified only during the colonial era (Mukharji 2016). At the same time, practitioners attempt to modernize Ayurveda by finding "genetic correlates" of *prakriti*. Ayurgenomics thus translates the ancient wisdom of Ayurveda into the modern language of commerce and genetics to produce a commodifiable system of personalized medicine and targeted pharmaceutical drugs (Prasher et al. 2017). While not without critics who call it "unscientific" and "pseudoscientific," the project has nonetheless received hefty government support, roping in dozens of researchers from leading institutions (Pulla 2014). "We are trying to contemporize Ayurveda," explains Mitali Mukerji, a molecular biologist who leads Ayurgenomics research at the Institute of Genomics and Integrative Biology in New Delhi (quoted in Pulla 2014). What we have witnessed is a fundamental revision of Ayurveda from a medical cosmology that embraced "mysticism, superhuman capacities, intangible agencies and mysterious therapeutic powers," into one that conforms to "predictability of standardized human biology, the calculable rationalities of population-level drug trials, and government policies, and the neat essentialisms of colonial identity politics" (Mukharji 2016). Quite a transformation indeed! Ultimately these projects secure the modernity of India through producing new medical commodities, and thus help consolidate bionationalism through genomics.

CONCLUSION: "UNITY IN DIVERSITY" AND GENOMIC NATIONALISM

> Memory has its own special kind [of truth]. It selects, eliminates, alters, exaggerates, minimizes, glorifies, and vilifies also; but in the end it creates its own reality, its heterogeneous but usually coherent version of events; and no sane human being ever trusts someone else's version more than his own.
>
> SALMAN RUSHDIE, *Midnight's Children*

Acts 1 and 2, one set in India and one in the United States, highlight seemingly contradictory claims—the first of a global "black" brotherhood and the other of an indigenous and unique India. Act 3 describes the creation of an Indian genetic database and the geneticization of Ayurveda, both best understood in their global contexts. Together they highlight the wide circulation and easy appropriation of genomics to very different political, biological, activist, national, and transnational ends. As Yulia Egorova (2013) argues, DNA testing can be empowering and disempowering, as it can "both reinforce reductionist accounts of human sociality and serve as rhetorical tools for social and political liberation." Several other initiatives also use genetics toward questions of identity and politics, such as those of Jews in India (Egorova 2006, 2013) and those of Parsis (Avestagenome Project). Bionationalism emerges as a unifying thread—the Indian state choosing to contain caste oppression within its national borders, diasporic Indians attempting to consolidate their multicultural pride in a unique and indigenous India, and the Indian state asserting its genomic sovereignty through the Indian Genome Variation Project and celebrating its indigenous heritage through geneticizing Ayurveda.

The proliferating studies present a paradox of sorts (Oikkonen 2014). On the one hand, these global genealogical studies are trying to produce global histories by erasing the geographic boundaries between the national and the global. Genetic ancestry, migration patterns, and demographic maps of diseases create global maps predicated on similar genetic technologies, methods, and analyses (Johnston and Thomas 2003). Yet at the same time we see the proliferation of nationalism and projects creating national genomes. These efforts simultaneously strengthen national genome projects that ground national histories in prehistory. This has allowed us to tell transnational, national, and nationalist histories at the same time. In developing a national history, genealogical data engage with the intimate politics of a nation's social and ethnic differences.

What genetic genealogical projects have unleashed is a plethora of nostalgic ruminations. Nostalgia is a potent narrative tool in imagining the evolutionary origins of the nation, allowing us to bridge the gap between national history and global prehistory through the cultural ideas of genetic knowledge (Oikkonen 2014). Sheila Jasanoff argues that modern science can often be circumscribed by nations (Jasanoff and Kim 2009; Jasanoff 2011). Here we see how science and religion cocreate the shape of the debates and the emerging frames of bionationalism. This is evident in the narration of the peopling of India, as well as in the nostalgia over India's glorious past that has long circulated in the flows of Hindu nationalism. The narration of history, the desire to belong, and the recovery and reconstruction of a homeland are very evident in modern Hindu nationalism. In the rise of Hindu nationalism in India, the globalization of Hinduism by Hindu nationalists in India, and the celebration and dissemination of Hinduism by diasporic Indians lies the emergence of a global/universal Hinduism.

The three sets of stories that animate this meditation on genomics in the twenty-first century in India tell a complex tale about modern Indian science in the making. National and diasporic communities are become deeply invested in such science, and teams routinely include researchers and samples from all across the globe. Contestations arise between religious nationalists and secularists, between historians and geneticists, between disciplinary methods and political confabulations, between global brotherhood and exclusive genetic citizenship, between shared ownership in the global domain and national "genomic sovereignties." By a swab of a cheek or a vial of blood, geneticists claim to track the telltale markers of nation, caste, race, and gender in our genome and generate an "authentic" history, and indeed an objective, scientific, accurate, and more definitive history.

While the rhetoric of these studies is cloaked in liberal discourses of secularism, multiculturalism, and antiracism, many scholars have documented that modern knowledge often reproduces old histories, reinscribing disputed biological categories such as gender, race, caste, and sexuality, as well as elite power structures (Skinner 2006; Egorova 2010b). Recent genetic technologies have thus opened up new spaces for challenges, possibilities, and aporias—biological, political, historical—that have reignited old debates. Global enterprises—including national governments, global organizations, private companies, for-profit and nonprofit groups—that shepherd this bourgeoning technoscience have a vested interest in the promises of defining and controlling not just biology or the body but rather "life itself" (Rose 2006).

What I hope seems obvious in this retelling is that new genetic technology gets embroiled in the old politics of race and caste as well as science and the humanities. New genomics appears to reinvigorate old categories of difference. What genetic testing does is shift social identity to genes and biology with little acknowledgment of their coproduction. Genetic technologies have moved out of research laboratories into public life. With them arise necessary debates about their proper use and interpretation (Rogers 2001; Brodwin 2002; Elliott and Brodwin 2002).

These are the tales of the secular and the theocratic, of science and religion, of the first and third worlds, of disembodiment and embodiment in the twenty-first century. No longer does science belong only to the West; the East and others claim it too. No longer is science confined to secular spaces and objective natural worlds; science has its own gods and rituals. No longer is science objective, saved from the tainted hands of politics, religion, and commerce; it is openly and deeply implicated in them all. No longer is science practiced only in the hallways of esteemed ivory towers; it is now democratized, claimed by all who want to, and widely practiced in various labs as well as garages and bedrooms across the globe through DIY science. No longer does the West control the knowledge or tools of production; nationalism has claimed genetics within its repertoire of tools. These are the new confabulations of technologies of belonging in the twenty-first century, the contours of a new biocitizenship in these neoliberal times.

AVATAR #5

The Story of Arul

Arul was the Kanu of grace, cultivating beauty, dignity, and refinement on the planet Kari. Its vegetal form flowed gracefully around it. From its purple central trunk, cascading roots and shoots emerged. Its gray roots spread across the billowy clouds as it constantly observed and updated the various avatars. Arul liked to believe that it shaped all the other avatars through its burrowing roots. Its sensory blue hair flowed off the top of a curvaceous trunk and fanned Avatara Lokam. Roots and shoots together, Arul was everywhere! Uruvam always marveled at Arul—its sinewy form, its graceful airs and intricate flows. In this new iteration, Arul became more bold and directed. It reveled in its recent innovations that would make it impossible to reduce its complex imaginations into boring, reductive formulations. Arul had built worlds within worlds. Scale was important—biological theories of molecules could not mimic those of species or groups. Each needed different considerations. Newtonian physics would have to work alongside quantum physics until the human species came to an advanced state of understanding life—scientific elegance alongside humanistic rigor, mathematical creativity alongside moral exactitude, artful biologies alongside nimble chemistries, playful philosophies alongside rigorous arts. Complex organisms emerged through millennia of evolution, playful symphonies of relations, exchange, and mutual engagement.

Homo sapiens used Arul's playful technologies to further their goals of uniformity, policing increasingly narrow conceptions of human life. No variation was tolerated, and the purity brigades wandered the streets to make sure the norms were strictly followed.

But as Arul watched, and Uruvam and Nādu cheered on, *Homo impuritus* finally put it together. They probed inside and out. As they probed organisms and their histories, they discovered they were not one, but many. They discovered that they embodied layered history of multiple coevolved organisms: organisms within organisms, cells within cells, genes within genes, all having come together from different places and times. All truth, all organisms, and even the laws that governed the planet were made of layers within infinite layers. Peeling away one would only reveal another below. *Homo sapiens*, in contrast, continued to prefer universal theories and politics of belonging. To them, time and place were specific. Everyone was "native" somewhere, and enforcing native and "pure" geographies and biologies was central to their imagination. *Homo sapiens* had no interest in peeling the layers of truth. They were happy to enforce their newfound discoveries and rule the planet with their purity doctrines.

Arul was particularly proud of its imaginations of life on Kari, creating infinite mechanisms by which organisms could live and produce new organisms. The essence was *change*—through recombination, gene exchange, syngenesis, dosage compensation, mutation, adaptation, plasticity, chromosomal exchanges, gene regulation, epigenomes, and a hefty dose of the goodness of random chance! Arul had ensured that patterns could be deceptive, that what initially appeared to be randomness or chaos might actually be very deterministic, and that what appeared simple might actually be very complex. Nothing was ever what it appeared. Arul's creation necessitated critical examination that invoked both rigor and play—a necessity Arul considered its most vital invention. This was the whole point of the evolutionary game, a vital aspect that even the powerful Kankavars could not fully control. Kankavars discovered that small acts of random chance could have large effects on the course of evolution, but these paths proved irreversible—hence the need for new iterations of life on Kari. While some assemblages of cells thought of themselves as discrete organisms, like *Homo sapiens*, many others enjoyed the playful versatility of life. Bacteria conjugated at will, slime molds came together in exuberant and playful formations, fungi carried on with no attention to demarcating individuals, plants communicated with each other through chemicals, primates spent endless time playing with and pleasuring each other. In contrast to *Homo sapiens*, who resolutely maintained species boundaries and a hierarchy of life, *Homo impuritus* embraced the fuzzy boundaries of species—with time they learned to communicate across species. They talked, loved, grieved, and related with plants, viruses, bacteria, rocks, oceans, animals, wind, fungi, and molecules to create new enclaves of rich and complex communities of life. In their exchanges with other species, they discovered preternatural and animate worlds— and soon they understood the existence of Avatara Lokam and the cycles of life and death.

With time, *Homo sapiens* were left behind, as *Homo impuritus* gained a deeper understanding of the Kankavars and their embedded clues in earthly biology, geology, and culture. Humans had discovered processes such as crossbreeding, genetically modifying organisms in the laboratory, transgenesis, regenesis, re-creation, and techniques like PCR, CRISPR, PLTPS, REPT, ALT, and CRT. It was obvious how the two human societies diverged. Most *Homo sapiens* used these technologies to strengthen nations and families and to bolster the boundaries of species and individuals. In contrast, *Homo impuritus* (and some *Homo sapiens* who joined them) embraced these technologies with play, desire, and pleasure, reveling in the endless possibilities that creation had enabled. Rather than create species, familial, geographic, ideological, or national boundaries, they used these technologies to explore the cosmic possibilities of their inheritance. Arul was delighted that most creatures on Kari joined *Homo impuritus*, and went about their many assorted and ebullient ways of life and living.

As new cross-species socialities emerged and evolved on Kari, life blossomed. The layered histories of time, scale, and place were embraced with new gusto. *Homo impuritus* realized that all existence shared in a common cosmology, and that greatest understanding came through collaboration. Geological scientists worked with literary theorists, mathematicians with artists, atmospheric scientists with bacteriologists. They unraveled the multiple layers of organisms, rocks, atmospheres, socialities, and stories. Digital landscapes merged with biological and literary landscapes. Through the fissures in time, one could travel through hyperlinks into other ideas, concepts, theories, biologies, and worlds. The material and immaterial, human and nonhuman, animal and plant, supernatural and natural, and morality, ethics, and justice were all woven into an expansive and porous cosmology. Arul was pleased. Finally, it thought, perhaps what it had worked so hard to create—grace, beauty, creativity, play, and dignity— might be here to stay. Only time would tell.

CHAPTER FIVE

Conceiving a Hindu Nation

(Re)Making the Indian Womb

Such is the power of the Indian soil that all women turn into
mothers here and all men remain immersed in the spirit of holy
childhood.

MAHESWATA DEVI, *Draupadi*

We were the people who were not in the papers. We lived in the blank
white spaces at the edges of print. It gave us more freedom. We lived
in the gaps between the stories.

MARGARET ATWOOD, *The Handmaid's Tale*

"A GOD TRANSFORMS INTO A NYMPH AND ENCHANTS ANOTHER
god. A king becomes pregnant. Another king has children who call him
'father' and 'mother.' A hero turns into a eunuch and wears female apparel.
A prince discovers on his wedding night that he is not a man. A princess
has to turn into a man before she can avenge her humiliation. Widows
of a king make love to conceive his children . . . these are some of the
tales I came upon in my study of Hindu lore," begins Pattanaik's (2002)
book on queer Hindu mythology. The sexual landscape in the Indian
imagination is expansive. In Hindu teachings, sex is not limited to procreative sex. Sex can have many purposes, all sanctioned (Pattanaik 2002).

It can be an act of dharma, a duty, an act with an express procreative purpose to create family. Sex can also be entirely about pleasure, or *kama*, and indeed pleasure and desire permeate Indian mythological storytelling. Sex can be a material transaction, toward *artha*, for commerce, or for producing heirs to thrones. And finally, it can also be an act of salvation, or *moksha*. In ideas prevalent in the *tantrik* texts, through sex one can gain magical powers and the ability to control nature. Sex can bring salvation, a break from the unending cycle of birth and death. Nonnormative sex and sexuality is also abundantly in display in Indian mythologies. Ardhanarishvar, a manifestation of Shiva and Parvati, is half man, half woman. In the *Mahabharata*, Gandhari gives birth to a lump of immovable flesh. This flesh is divided into a hundred and one pieces, and each is put into a nutritive jar and incubated. From each, a child is born. Parthogenesis, or asexual reproduction, is rampant—individuals become pregnant through the elemental gods, through chanting unique mantras, or purely through desire. Sex and gender are fluid, as characters can change their sex and gender (including human-animal hybrids). Reproduction is decoupled from sexuality or sex, as a plethora of imaginative possibilities unfold in these stories. Indian mythology can trouble gender and sexuality in its delightfully playful plots and imaginative possibilities, offering endless permutations and combinations of the sexual.

In this chapter, I want to continue to explore India as an archaic modernity by tracing the history of the "womb" as a site of politics. As Janaki Nair and Mary John (2000) argue, women are "reproductive beings," and control over their bodies has proved to be a central site of political action. Brahminic patriarchy controlled the "purity" of upper-caste women (U. Chakravarti 1993). The dangerous sexuality of the "non-mother" motivated social reform legislations of the nineteenth century, and the stigmatization of the poor as irresponsible and promiscuous prompted national programs to control their fertility (Tambe 2009). As women became the symbol of Indian "authenticity" and a sign of the superiority of the East, the bifurcated invocations of womanhood and femininity emerged. The entrenched politics of caste and gender add more complexity. Feminist movements in India have long grappled with celebrating caste-based occupations and the skills of health and healing of their practitioners such as midwives, even as they are pushed by modernist discourses to delegitimize caste-based occupations (Gopal 2017). Hindu nationalists promote a vision of strong women through the figure of the female goddess, or Shakti, but at the same time insist that women's strength is best served through their role in the family and the nation. Here, I trace how the womb has long been a central focus for feminists. What emerges is an evolving technopatriarchal, transnational morality play. This play moves us from earlier narratives of third-world hyperfertility, in the claims of overpopulation and population explosion that "control" women's bodies and sexuality through long-term contraception and sterilization, to the new clinics of technoscientific surrogacy where the womb is again controlled— as an entrepreneurial subject through gestational surrogacy, through paternalistic laws outlawing commercial surrogacy, or through recent eugenic attempts to revive ancient Vedic gestational sciences. In each of these scenes of the morality play, women's bodies become sites of bio-political control.

My interest in surrogacy emerged with the rise of Hindu nationalism and its attendant tensions. Gujarat emerged as a surrogacy outsourcing capital of India (Nayak 2014) during the years when the current prime minister was its chief minister (*Daily Bhaskar* 2013). Estimates suggest that in 2013, 40 percent of surrogacies in India were from Gujarat (Jayaraman 2013; *Daily Bhaskar* 2013). Government regulation, or rather a singular lack of regulation, is key to the success of the surrogacy industry (Mahapatra 2008; Bhalla and Thapliyal 2013). Some observers characterize the growth of surrogacy in Gujarat as an accomplishment of Modi's "development model" (Austa 2014). To others, surrogate mothers emerged as a symbol of "women's empowerment" and a site of "entrepreneurship" (Bedi 2013; Rabinowitz 2016). The *Gujarat*, a quarterly magazine published by the state Commissionerate of Information, argues that "the state has set a precedent in embracing humanist ideas by facilitating reproductive tourism, which has proved immensely useful. . . . Apart from empowering the surrogates, it is bringing a lot of revenue for the state itself, furthering its development" (Nair 2012). A Hindu nationalist–led state government nurtured a multibillion-dollar industry, arguing that India's grand Vedic civilization would through the surrogacy industry once again become the new "cradle" of world civilization (see Lee 2014). Melding a rhetoric of entrepreneurship with claims of scientific and medical excellence, surrogacy has made deep inroads as several generations of women in the same family are often recruited or volunteer to serve as surrogates (Yagnik 2013). Surrogate bodies are "potentialized" through a combination of surrogacy and networks of social and economic inequality (Vora 2013). The state has been important in crafting a "manufacturing model" that makes this a commercially viable enterprise, efficient in time and money.

Clinic doctors like Dr. Nayana Patel, now world famous through appearances in TED talks, television programs such as *HardTalk* and the *Oprah Winfrey Show,* and documentaries such as *Womb of the World* and *House of Surrogates,* was honored by Chief Minister Narendra Modi at the district Mahila Sammelan, or women's conference.[1] The success of gestational surrogacy emerged alongside the touted "Gujarat development model," a highly successful platform that brought Modi and his Hindu nationalist party to national victory in 2014. How did a paternalistic Hindu nationalism, deeply grounded in a muscular masculinity, become a site for gestational surrogacy? As critical analyses dubbed the Gujarat development model "unregulated neoliberalism" (Hensman 2014), the embrace of surrogacy became clearer. The central biopolitical nature of Hindu nationalism and its belief systems also became apparent in long-established beliefs in divine intervention that encouraged and enabled surrogacy through "surro-gods" (Pande 2014a).

The neoliberalism of Gujarat was replaced by a more familiar paternalism in 2016, when the central government banned all commercial surrogacy in order to "protect" women. Also emerging were new models of Hindu "Vedic" wisdom for pregnancy that promise eugenic visions, straight out of Nazi playbooks, of "taller, fairer, and smarter babies" (Gowen 2017).

BIONATIONALISM: (RE)BIRTH OF THE CLINIC

> Even Ecuadorian bio-scientific practitioners do not imagine their scientific endeavors as "surpassing" a religious past; instead, they live in a fully religious present.
>
> ELIZABETH ROBERTS, *God's Laboratory*

In *The Birth of the Clinic*, Michel Foucault (1994) theorizes that the development of modern medicine and its experts is tied up with the politics and emergence of new forms of governmentality, the management of social and political spaces, and ultimately the regulation and disciplining of biological bodies. I revisit the idea of the "clinic" and move it across time and place by examining the histories of the "pregnant" body in postcolonial, independent India.

Campbell (1992: 35) argues that "contemporary medicine has transformed the human body into a source of instrumental value, a resource of value to others: patients, physicians, and researchers. . . . Such practices seem to presuppose a basic feature of property, that is, the capacity and power of alienation or transfer." Indeed, gestational surrogacy disaggregates women's bodies as resources, rendering the womb as a disembodied, "empty," and "not being used" resource that is available to make money. Like sperm, ova, and organs, wombs have also been isolated as an individual commodity (Nayak 2014). The body is entirely abstracted and commodified and transformed into a "manufacturing mode" of (re)production (Darling 2014). As Kalindi Vora (2015b) has argued, the "technofantasy of the isolated womb" allows for the conditions of the structure and discourse of Indian surrogacy as "wombs for rent," diminishing any sense of societal responsibility for the surrogate mother's life other than during the period of the surrogacy contract.

PREGNANT POLITICS AND TECHNOPATRIARCHY: THREE ACTS

The pregnant body and the Indian womb are a dynamic site of politics. In many ways, the modern surrogate mother living and working in postcolonial India follows earlier figures of the *amah* and the *dhayelayah*, the Indian wet nurses and nannies of the empire (Jacob 2015). Since India's independence from British colonialism, the pregnant body is the icon of global imaginaries of India, through secular and Hindu nationalist governments. One can characterize three distinct phases of reproductive politics, which I discuss in three acts. The first is a prelude to the rise of recent Hindu nationalism, and the other two describe the evolving politics of Hindu nationalism in contemporary India.

Act 1. Prolific Pregnancies: The Overpopulating Womb

Since Malthus's famous 1798 treatise *An Essay on the Principle of Population*, which postulated a "struggle in society" resulting from scarce resources in the face of growing populations, "Malthusian logic" has never left us. Eugenic logic and the control of women's fertility began in India during British colonial rule. Eugenic scripts (Subramaniam 2014) and the eugenic logic of controlling women's fertility were embraced in India during colonial rule. Eugenics, which emerged as a scientific way to modernize India, was adopted by Indian nationalists and British colonizers alike (Hodges 2008, 2010). In postcolonial India, Malthusian logic has been updated with new scientific and economic realities in "neo-Malthusian" logics. This is extremely well researched ground in feminist circles, and feminists have powerfully demonstrated that the primary targets and victims of this logic have been the poor and women of color in the West and women in the third world (Hartmann 1995, 2017; Williams 2014; Murphy 2017). Curtailing the reproductive capacity of the third world through the long-enduring figure of the "hyperfertile" third-world woman is well documented in feminist theory and activism. The rhetoric of overpopulation is powerful. Despite the critical work in feminism, the idea of overpopulation continues to have supporters even in feminist circles; Adele Clarke and Donna Haraway's recent work is a case in point (Haraway 2015, 2016; Clarke and Haraway 2018).

Since the Population Council identified India as the location where the population problem was particularly severe, India has been ground zero for the politics of overpopulation (Williams 2014). Carole McCann's (2016) important study *Figuring the Population Bomb* gives us a chilling analysis of how faulty reasoning and a "fictitious accuracy" in numbers kept unrelenting claims of Indian overpopulation firmly in place. The Indian womb as prolific and overpopulating is by now legend, and circulating logics of women's grinding poverty, poor education, Indian patriarchal culture, lack of reproductive control over their bodies, and poor status in Indian society have emerged as widely circulated reasons for the passivity of Indian women. As a result, and with the full cooperation of the Indian state, India has been the world's laboratory for birth control, long-term contraception, coercive family planning, and sterilization. Overt coercive state policies during the Emergency years (1975–77), including forced vasectomy programs, are a startling prologue to the rise of the surrogacy industry a few decades later.[2]

As Hindu nationalism has grown, the alleged hyperfertility of Muslim women has been highlighted as a site of national anxiety. In the 2017 state elections, the prime minister, while campaigning in Gujarat, alluded to the idea of the overfertile Muslim family. He repeated the slogan from a 2002 election speech, in which he alleged that Muslim families practice the productive policy of *"hum panch, hamara pachees"* (we five, ours twenty-five), insinuating a Muslim polygamous husband and his four wives, and their twenty-five children (R. Guha 2017; Salam 2017). The hyperfertile Muslim family and the overpopulating poor both prove to be powerful tropes that run through postcolonial India.

Act 2. Pecuniary Pregnancies: Wombs for Hire

As assisted reproductive technologies developed, India's fertility emerged as rich terrain for technological intervention—ultrasound (especially for sex selection), in vitro fertilization (IVF), and other technologies of assisted reproduction have proliferated (Aditya Bharadwaj 2006a, 2006b; Bhatia 2017). As technologies of surrogacy evolved, commercial gestational surrogacy grew steadily.

The term *surrogate* is derived from the Latin *subrogare*, which means "appointed to act in the place of." Gestational surrogacy involves implanting an embryo created through IVF into a surrogate mother who carries the fetus to term. In contrast to traditional or genetic surrogacy, in which the surrogate is the baby's genetic mother, a gestational surrogate mother does not contribute genetic material to the offspring. While commercial surrogacy is illegal in many parts of the world, it was until recently a burgeoning industry in India (Sama 2012a, 2012b), largely unregulated since inception and then briefly regulated (A. Menon 2012). Indeed, India soon emerged as the "mother destination" for gestational surrogacy (Rudrappa 2010). Here we need to understand the postcolonial Indian womb as an important site of analysis. The emergence of the Indian womb as the "mother" of destinations is a confluence of several factors: India's embrace of modern technology has created a proliferating market for "medical tourism"; a strong workforce of doctors and technicians have provided the technical expertise; the continuing inequality within India has consigned a vast section of the population to low-tech and marginal living, with little hope of upward mobility, and in them has created disciplined workers; and the proliferation of globalized and neoliberal privatized markets have provided the contexts. A scientized religion melds religion and science into a palatable commodification of the womb, and a religionized science translates Western science and technology into desirable commercialized pregnancies. As a result, commercial gestational surrogacy in India grew into a multibillion-dollar industry—from USD $445 million in 2008 to over USD $20 billion in 2011 (Nayak 2014). Technoscientific surrogacy typically employs high-tech reproductive technologies and a low-tech and economically marginalized workforce (Goodman 2008).

Through secular governments and even certain Hindu nationalist state governments like Gujarat, gestational surrogacy grew. India emerged as a site of "reproductive travel," where infertile couples from India and the Indian diaspora as well as non-Indians from abroad have come to India to "rent a womb" for their potential embryos from a gestational surrogate (S. Carney 2010; Voigt, Kapur, and Cook 2013). This is a global industry with complex and multiple circuits of travel, where intended parents enter into "reproductive exile" to go to another country for conceiving a child. The circuits are so complex and transnational that some Indian couples are priced out of India and have to travel to foreign countries like Dubai for various forms of reproductive technologies (Inhorn 2012).

Gestational surrogacy in India is by now a well-theorized site in feminist politics.[3] Together, scholars have raised important patterns: how the womb has emerged as a technologically mediated organ and been rendered a disembodied, desexualized site of labor; how the politics of gender, race, class, caste, and religion have made the wombs of some bodies more "bioavailable" than others; how patterns have revealed the reproductive stratification across the first and third worlds (Pande 2010; A. Banerjee 2011; Vora 2015a) as well as within the third world; how the antinatalist politics of the population bomb have now been transformed into pronatalist through gestational surrogacy; how surrogacy has been entirely medicalized, and how women who are the most vulnerable and precarious in their health and wealth emerge as the key site for surrogacy; and how lax, client-friendly regulation endanger the surrogate mother's body. A burgeoning number of academic and journalistic accounts have chronicled complex and fascinating narratives of surrogacy and the experiences of surrogate mothers. Indeed, various towns in India, such as Anand in Gujarat—referred to as the surrogacy outsourcing capital of the world—have become famous for their surrogacy centers, where surrogate mothers live in hostels for the length of their pregnancy, closely surveilled and monitored for optimum fetal development (Voigt, Kapur, and Cook 2013).

Prabha Kotiswaran (2018) outlines three phases in the regulation of surrogacy. First, a medico-liberal phase between the 1990s and 2008 liberally permitted the commodification of reproductive labor for the market. A second, contested phase from 2008 to 2012 incorporated regulations that ensured rights for the surrogate mother, expanded who could access surrogacy, and imposed obligations on commissioning parents. The third, a contracting and normative phase of 2012–2017, attempted to further regulate surrogacy, culminating in a ban on all commercial surrogacy in 2016. However, deep differences remain even within the government over how surrogacy should be regulated.

There is little doubt that gestational surrogacy through biomedicine has deep eugenic roots (Pande 2014a; Deomampo 2016). Potential surrogate mothers and egg donors with caste and class privilege command a huge monetary premium and greater negotiating power in the marketplace (Nayak 2014; Bailey 2014). Indeed, biological matter from poor women, women of lower castes or minority religions, and those with disabilities always proves to be at a disadvantage. Anecdotal evidence suggests that some parents prefer fair-skinned, educated, and upper-caste women, whether out of "eugenic" claims or because they feel that such women are more hygienic (Sama 2012a). Despite these preferences on the part of hopeful parents, the data suggests that the primary reason for becoming a surrogate mother is economic necessity. As Saravanan (2010: 27) argues, many surrogate mothers are submissive to the demands of doctors and intended parents, and often they are "on the edge of poverty" because of indebtedness, homelessness, and lack of education and remunerative work. Other studies find surrogate mothers to be professionals, albeit facing many layers of dominance and power over their lives (Vora 2015a).

While there is variation across India, a dominant narrative of gestational surrogacy has emerged. The debates, as Susan Markens (2012) argues, have revolved around the question of whether the globalization of reproductive labor is an exploitation of the surrogate mother or an opportunity for her and, in a related vein, whether the surrogacy narrative is best understood as gendered altruism or gendered empowerment. As A. Banerjee (2011) argues, the literature largely presents this debate as an ethical issue of reproductive liberalism versus exploitation. Others, like Bailey (2011, 2014), have attempted to move beyond the binary to create new ethical frameworks for feminist reproductive justice.

All the above factors are critical to explaining the emergence and rise of gestational surrogacy in India as a multibillion-dollar industry. In drawing on the biopolitics of Hindu nationalism, I am not attempting to compete with or supplant these important analyses.

As the neoliberal privatized model of gestational surrogacy exploded, the exploitative conditions of surrogacy became apparent. Feminist activists have actively uncovered these conditions: inadequate payment (with doctors and agencies keeping the lion's share), lack of health insurance, and lack of support or recourse after the pregnancy if there is a miscarriage or if health issues emerge after the birth. Surrogates have often signed legal contracts without understanding them, and allegations of captive women in tightly guarded guesthouses were widely reported (Lakshmi 2016). After several unfortunate cases, it was clear that something needed to be done. For example, as recently as 2017, highly publicized surrogacy rackets were discovered in Hyderabad (*Deccan Chronicle* 2017) and Bhongir (Nadimpally, Venkatachalam, and Raveendran 2017). High-profile cases— Baby Manji Yamada in 2008 (Mahapatra 2008) was abandoned by the commissioning parents after a divorce; an Australian couple abandoned a boy child in India and left with their daughter (J. Carney 2015); a British family was initially unable to get a British passport for their child conceived through surrogacy because Britain does not recognize commercial surrogacy (Neelakantan and Ganesh Kumar 2016)—all demonstrated the need for more stringent laws. With a Hindu nationalist government at the helm in 2014, this area was ripe for a new act in our morality play.

Several aspects of gestational surrogacy are worth emphasizing. First, the demography of surrogate mothers shows deep social stratification of those who are hired. Although primarily motivated by economics, they span the poor and the lower middle class. Second, caste discrimination in India continues in all walks of life and is often correlated with class. In orthodox settings, upper castes are not allowed to even touch lower castes, use the same cups or plates, or share the same public or private spaces. Such practices continue even to this day where critical utilities like wells, often the sole source of water in certain villages, are inaccessible to those of lower castes, as are temples and other sacred areas of worship. In a country where Hinduism is deeply entrenched in the politics of purity and pollution, it is astonishing how intimate practices such as gestational surrogacy still carry on across caste and class lines. How do upper-caste couples rationalize contracting a surrogate mother from a lower caste? The data would suggest that in practice, parents looking for a gestational surrogate mother ignore the strictures of caste and class, often because of cost and availability of surrogate mothers. Sometimes, requests for "vegetarian" surrogate mothers were really a code, a stand-in for caste concerns.[4] Third, despite the medicalized mode of surrogacy, there is little focus on the health history of the surrogate mother or any concern over epigenetic factors that have grown in importance in the rest of biology. It is striking, as Deboleena Roy (2017) has argued, that the cheapest surrogate mothers are from Bhopal, the very bodies still living out the effects of the gas leak twenty-five years ago. Finally, as some scholars have suggested in other national and transnational contexts (S. Kahn 2000; E. Roberts 2016), the scientific and medicalized processes of surrogacy are modulated by religion and by local spiritual practices.

Act 3. Persecuted Pregnancies: Protecting the Womb

As Ketu Katrak (2006: 213) reminds us, "Patriarchal ideology, in supervaluing motherhood, paradoxically contains and controls women." In August 2016, the BJP government proposed a complete ban on commercial surrogacy with the Surrogacy (Regulation) Bill 2016. The draft bill was passed by the Union cabinet on August 24, 2016, and then it was sent to Parliament (P. Bharadwaj 2016). In banning surrogacy, Hindu nationalists exercised their biopolitical power in reconsolidating the centrality of the heterosexual family. The proposed law sought to "protect women from exploitation and ensure rights of the child born through surrogacy" (Dhar 2016). The bill allows altruistic surrogacy for "needy infertile couples," but access is restricted. Only Indian couples (specified as man and woman) who are citizens, have been legally married for at least five years, and produce a certificate from a doctor saying they are medically unfit to produce a child can enter into an altruistic agreement. Eligible women should be 23–50 years of age, and men 26–55 years. To avoid "commercial exploitation," the surrogate mother must be a "close" relative (like a sister or sister-in-law) who is married and has at least one healthy biological child. A woman can be a surrogate mother only once in her lifetime. There can be no exchange of money between prospective parents and the surrogate mother. The only expenses allowed are for medical bills, "which will be paid to the clinic."

Those who cannot use surrogacy include foreigners—even diasporic Indians who are nonresident or hold a Persons of Indian Origin card. The law also prohibits Indian citizens who are "homosexual," "single," or "live-in couples." Couples who already have children, biological or adopted, cannot avail themselves of altruistic surrogacy. The bill proposed a hefty penalty of imprisonment of not less than ten years and a fine of 1 million rupees.

Particularly revealing were the reasons given by the government for the bill's specifics. Repeatedly, in answers to numerous questions, the government identified the infertile (but decidedly sexually active!) married heterosexual couple as the sole individuals deserving progeny. All others were undeserving because they were deemed Westernized, too fashionable, too casual about children, or not belonging to "a true Hindu ethos." Sushma Swaraj, the external affairs minister, specifically pointed to several high-profile cases of surrogacy by proclaiming, "Surrogacy is not a fashion or a hobby, but we have surrogacy as a celebrity culture" (Dhar 2016). What began as a convenience has become a "luxury today," she argued (*Express Web Desk* 2016). When asked about the exclusion of homosexuals, Swaraj said it was appropriate since homosexuality "is against our ethos." The reason for barring nonresident Indians and foreigners was a concern for the child because "divorces are very common in foreign countries" (Lakshmi 2016). Banning commercial surrogacy was the duty of the state and a "revolutionary step for women's welfare" (Lakshmi 2016).

Thus far, the courts had largely championed only the rights of the commissioning parents by reaffirming parenthood as strictly within the confines of genes and heredity (P. Bharadwaj 2016). Some clearly celebrated the government's ban, but its critics were also swift, numerous, and varied. While some feminist groups had always called for a ban, they found the distinction between commercial and altruistic surrogacy problematic. Activists who supported commercial surrogacy had long advocated for stricter regulation (Sama 2012a). Others raised even more fundamentally troubling issues (P. Bharadwaj 2016; Dhar 2016; Nadimpally, Venkatachalam, and Raveendran 2017; S. Sharma 2017): Why was surrogacy linked to marriage? Would banning commercial surrogacy rather than regulating it only drive it underground, causing more exploitation than an outright ban? The highly publicized surrogacy rackets in Hyderabad clearly show thriving black markets. Why would a ban help? Who would regulate "close" relatives? In familial situations in a patriarchal culture, what does it mean to "consent"? Can altruism within patriarchal structures be free of coercion? Why should reproductive labor be so undervalued as to not merit payment? Might people present commercial surrogate mothers as altruistic while paying them under the table? Other than criminalizing surrogacy, there was nothing in the law that kept surrogacy in check. The law had paid no attention to surrogate mothers; there were no safeguards of any kind, making the surrogate "another process in the chain of (re) production" (P. Bharadwaj 2016).

On August 10, 2017, a parliamentary report on the 2016 surrogacy bill was presented (Parliament of India 2017). The report itself echoed several of the issues the critics had raised. They argued that the expectation that women give their reproductive labor for free was "grossly unfair and arbitrary," and that forcing women into altruistic modes was "tantamount to another form of exploitation." They conjectured that individuals might be forced or coerced, and the results could be "more exploitative than commercial surrogacy." They also suggested that surrogate mothers should be entitled to health and life insurance. They were critical of the narrow conception of deserving couples and suggested that the state allow surrogacy for unmarried couples, widows, and divorced women; however, they did not include gay individuals or couples. Finally, they suggested a "compensated surrogacy model." As Nadimpally, Venkatachalam, and Raveendran (2017) argue, troubling issues remained. While the report suggests compensation, it was in practice more "discursive" than material as the surrogate mother could be compensated only for lost wages and medical care for mother and child. For a surrogate population that was largely poor and working sporadically in the informal sector, lost wages could be incredibly low.

In December 2018 the Lok Sabha passed the surrogacy bill. Only time will tell how this morality play will end, and whether women will ever have a hand in writing the play.

ARCHAIC MODERNITY AND ITS TECHNICS

> The laboratory has become one modern secular realm that is commonly thought to demand the explicit exclusion of spirit in order for the work of science to proceed. Its daily operations are thought to be "secular" in the extreme sense, signifying the full excision of religion from its domain. It is sometimes easy to forget, however, that the secular was initially constituted through the creation of religion as a bounded object and its subsequent constriction to the "private" sphere (another newly formulated category of the Enlightenment).
>
> ELIZABETH ROBERTS, *God's Laboratory*

The rise of Hindu nationalism and its vision of archaic modernities have ushered in deeply biopolitical projects. In reconsolidating the centrality of the heterosexual family and in banning all but "altruistic surrogacy," Hindu nationalists exercise their biopower. In thinking more broadly about Hindu nationalism and science, it is important to contextualize the role of religion and science in India. Religion has always been a central force in South Asia. As Ashis Nandy (2002: 139) argues:

> When gods and goddesses enter human life in South Asia . . . they enter human life to provide a quasi-human, sacral presence, to balance the powerful forces of desacralization in human relationships, vocations and perceptions of nature. This familiarity has bred not

contempt . . . but a certain self-confidence vis-à-vis deities. God and humans are not distant from each other. . . . Maybe this is the reason why allegiance to a deity is so often personal and looks like a bilateral contract or a secret intimacy between two unequal but sovereign individuals.

Gods and goddesses, alive and well in contemporary India, pervade everyday life. Lucinda Ramberg's ethnography of young Dalit girls given to the goddess Yellamma reminds us of the tremendous power of divinity in India. The goddess "has the power to afflict as well as the power to cure" (Ramberg 2014: 83). The goddess confers fertility, health, and wealth on those who please her, but she can cause misfortune to individuals and entire communities if not adequately cared for. We have seen a proliferation of god-men and god-women, an increase in claims of the divine benefits of products such as sacred stones, necklaces, amulets, and bracelets, and perennial pilgrimages to sacred sites and temples (Nanda 2011). Temples and ashrams, and their attendant gurus and swamis, have amassed great wealth, emerging as new sites of philanthropy in India (Denyer 2011). Religion and science emerge as deeply entangled (S. Kahn 2000; Aditya Bharadwaj 2006a; E. Roberts 2012). The pressures to procreate and for women to be mothers have shaped complex relationships between reproductive technologies and religions across the globe (Aditya Bharadwaj 2006b; D. Roberts 2009; S. Kahn 2000). To understand the deeply entangled roles of religion and science in India, and to understand how archaic modernities work, we need to examine two sites. One is the site of traditional biomedicine, and how religious practices in India continually subvert and supplement the "rational" practices of reproductive biology and biomedicine. The other is the proliferation of Vedic sciences and practices, and how they have engendered new eugenic imaginations of Hindu nationalism.

ENABLING BIOMEDICINE THROUGH RELIGION

A clinical theodicy. . . assists in making sense of biomedical uncertainty so as to resolve tentative outcomes of conception technologies. This is perhaps indicative of a larger process of double entrenchment of tradition in India today, a coming-together of reconfigured traditional

notions of divinity and the modern idea of clinical space. More cru-
cially, the indigenization of biomedicine in India demonstrates that
the clinical–biomedical modalities are far more open to uncertainty
and amenable to pluralistic understandings than the "technocratic
imperative" . . . permeating clinical application of science in the Euro-
American contexts.

ADITYA BHARADWAJ, "Sacred Conceptions: Clinical Theodicies,
Uncertain Science, and Technologies of Procreation in India"

Religion in India, in particular, demonstrates deep entanglements with
science (Aditya Bharadwaj 2006b; Kleinman 2016). Aditya Bharadwaj's
(2006b) wonderful ethnography of IVF clinics in India shows how religion
is everywhere, deeply imbricated in the practices of the clinics, the doctors, and their patients. Rather than adhere to ideological commitments
to "scientific rationality," we find that the divine, the spiritual, and religious beliefs fill "the vacuum of uncertainty" that underlies technoscientific practices. The doctors, Bharadwaj argues, are deeply committed to
induce conception in the IVF clinics, and they use their considerable cultural beliefs to provide successful outcomes. For example, in an interview,
a doctor explains that assisted reproductive technologies are an "incomplete science." Success rates are not very high. As someone who believes
in God, astrology, palmistry, and Vaastushastra, the doctor uses the power
of these alternate medical practices to improve the chances of success. He
sends his patients to homeopathy, Vaastu practitioners, and astrologers.
He argues that with the aid of these alternate religious and spiritual
practices, his success rate has increased. The incomplete science and its
attendant uncertainties are thus made complete by the assistance of "non-
scientific" astrology, Vaastu, homeopathy, and other alternate practices. As
Aditya Bharadwaj (2006b: 458) concludes:

> His predominant concern is to induce conception rather than to
> adhere to a broader ideological commitment to "scientific rational-
> ity." The eclectic combination of the clinical and the nonclinical,
> therefore, helps the doctor to manage the questions that biomedi-
> cine leaves unanswered. His spiritual beliefs, in this respect, stand
> in to fill the vacuum of uncertainty that clinical medicine has failed

to fill with so-called rational explanations. The engagement with divine or cosmic forces in this respect becomes a mechanism to make sense of the unknown in the working of the human body.

As Elizabeth Roberts (2012) reminds us in her study of assisted reproduction in Ecuador, the scientific laboratory is imagined as a secular realm that explicitly excludes the spirit in order for the work of science to proceed. She argues that it is worth remembering that the secular of the Enlightenment was "constituted through the creation of religion as a bounded object and its subsequent constriction to the 'private' sphere" (E. Roberts 2012: 512). Instead, what Roberts finds is that "nature is always in process—entangled in assisted relations with deities and humans, not fixed or outside of the human" (Sacco and Agro 2017). Like Ecuador, India too reveals porous boundaries between the scientific and religious. Focusing on surrogacy clinics in particular, Amrita Pande (2014a), in her excellent exploration of surrogacy in *Wombs in Labor*, demonstrates that the divine pervades the lives of surrogate mothers as well. She argues that women understand their work largely in the language of gods, goddesses, and the divine. Stories from Hindu mythology are used to argue that surrogacy is not new and is in fact part of the Hindu religion (Pande 2009b). While I disagree with Pande that there is an easy separation between organized religion and the divine of the everyday, the pervasiveness of the divine is striking. Fictional gods of surrogacy have emerged; surrogate mothers invoke "surrodev" and "surro-god." The doctors are seen as divine, endowed with godlike powers, and their words and advice are sanctified. Unlike narratives of surrogacy in other parts of the world, where surrogate mothers see themselves as angels and "gift givers" of life, Pande argues that surrogate mothers in India see themselves as blessed by the surro-gods to earn money to help their families and communities.[5]

This pervasiveness of divinity and gods sanctifies a surrogacy process that is thoroughly medicalized and commercialized. Despite the ethos of divinity, research shows that surrogacy is largely an outgrowth of economic necessity, undertaken by the economically underprivileged and those in need of money (*Hindustan Times* 2009; Kharandikar et al. 2014). The women are often illiterate and in a patriarchal setup where men are still in control, and the surrogacy process often leaves them with little agency in the contractual process (Pande 2010).[6] Narratives of divine intervention and disembodied narratives of the "womb" also enable a desexualizing of surrogacy, making it more palatable as "respectable" work. What is also striking is how thoroughly medicalized the process of surrogacy has become (Vora 2014). Scholars have outlined how the conditions of gestational surrogacy—living conditions, diet, lifestyle—have been entirely crafted with a scientific patina. Understanding the interconnected history of science and religion in India, where religion and science meld into each other, where doctors are gods, where surrogacy has its surrogods, and where surrogacy is seen as a boon or blessing by these gods, demonstrates the power of an archaic modernity, a scientized religion, and a religionized science.

NEW BIOPOLITICAL IMAGINATIONS OF HINDU NATIONALISM

One of the central and ongoing projects that Hindu nationalists have embarked on is to "Hinduize" the nation. Since coming to power in the national government in 2014, they have poured considerable investment into these projects—taking over research institutions, rewriting school textbooks and curricula, and reshaping research and policy agendas. In the realm of biology, Hindu nationalists have sought to modernize and scientize Vedic sciences by reconstructing them in the language of modern genomics. The new Ministry of Ayurveda, Yoga and Naturopathy, Unani, Siddha, and Homeopathy (AYUSH) has a separate budget and a higher status than any similar agency has ever had in India's history. On its website, images of Prime Minister Modi in yoga *asanas* with a large group of followers fill the screen. Over the last year, several efforts at reproductive enhancements using Vedic and Hindu sciences have garnered international attention. These practices and claims have an older history. For example, Lucia Savary (2014) describes what she calls "vernacular eugenics" in India during colonial rule in the early decades of the twentieth century. Known as *santati-śāstra* (the "science of progeny" or "progeniology"), this emerging branch of knowledge bases its principles on Francis Galton's "classical eugenics" but has adapted them to Indian eugenics, using Ayurveda or *ratiśāstra* (ancient texts that deal with conjugal love) as its knowledge base. During this time period, Savary (2014: 381) argues, "western science functioned as a legitimizing source in vernacular texts." In the more recent projects, what is striking and alarming is the seamless melding of the ancient and the modern to reconfigure Vedic medicine as proven knowledge.

Let's consider claims of the revival of the ancient Indian tradition of *garbh sanskar*, or education in the womb. Organizers claim that it "is a scientifically proven fact" and "an amazing way of teaching/educating and bonding with unborn baby in womb during pregnancy." Its objective is to produce *uttam santati*, superior children (Sampath 2017). Parents are advised to follow "three months of *'shuddhikaran* (purification)' for parents, intercourse at a time decided by planetary configurations, complete abstinence after the baby is conceived, and procedural and dietary regulations" (Ashutosh Bharadwaj 2017). Ashutosh Bharadwaj (2017) quotes a doctor as stating, "The shastras prescribe a specific time to have intercourse for pregnancy. Doctors tell couples when they should become intimate on the basis of their horoscope and planetary configurations." The program involves "purification of the energy channels" (Gowen 2017) and following the religious scriptures. This project, launched in Gujarat a decade ago, has been promoted at the national level since 2015. Its national convener, Dr. Karishma Narwani, states, "Our main objective is to make a samarth Bharat (strong India) through uttam santati [superior offspring]. Our target is to have thousands of such babies by 2020" (Ashutosh Bharadwaj 2017). Such training extends to Hindu nationalist camps called Arogya Bharati (Gowen 2017).

Through Ayurveda, the program argues, you can produce superior offspring: "The parents may have lower IQ, with a poor educational background, but their baby can be extremely bright. If the proper procedure is followed, babies of dark-skinned parents with lesser height can have fair complexion and grow taller" (R. Mishra 2017). A perfect example of an archaic modernity, the claims originate in both Indian mythology and modern biology. *Garbh sanskar* (education in the womb), for example, draws on the Indian mythological tale of Abhimanyu. In the *Mahabharata*, Abhimanyu is described as having learned the art of breaking the "chakravyuh" (a circular trap) inside his mother's womb as his father narrated the method (Ashutosh Bharadwaj 2017). Alongside this mythological insight, the project introduces bioscientific language. Repeatedly in the numerous projects that have proliferated, one sees the mingling of Indian mythological stories alongside bioscientific language that often proves nonsensical if one examines it carefully. For example:

> *Garbh sanskar* enables "genetic engineering in vivo or inside the womb." (*Indiatimes* 2017)
> This procedure "repairs genes" by ensuring that genetic defects are not passed on to babies. (Ashutosh Bharadwaj 2017)
> Ayurveda has all the details about how we can get the desired physical and mental qualities of babies. IQ is developed during the sixth month of pregnancy. If the mother undergoes specific procedures, like what to eat, listen and read, the desired IQ can be achieved. Thus, we can get a desired, customised baby. (Ashutosh Bharadwaj 2017)

Often the personnel have a mixture of traditional and bioscientific training. For example: "Narwani and Jani hold Bachelor's degrees in ayurveda, medicine and surgery, and Varshney obtained a PhD in biochemistry from Allahabad University in 1986" (Ashutosh Bharadwaj 2017).

All of the projects share a few features. The primary advice seems to be the control of the pregnant woman—making her a happy, docile, accommodating individual. They promote being "good" and religious, reading religious scriptures, listening to the *Ramayana*, and following austere Hindu values such as eating vegetarian food. The recent trend of violence against meat eaters is significant given that vegetarianism is a cultural practice of only a minority of India's population (Natrajan and Jacob 2018). The advice is decidedly puritanical in its prohibitions against desire and passion (albeit not against sex!).

> Pregnant women have been advised to stay away from "desire or lust", avoid non-vegetarian food and have spiritual thoughts. . . .
> Pregnant women should detach themselves from desire, anger, attachment, hatredness [*sic*], and lust. Avoid bad company and be with good people in stable and peaceful condition always. . . .

The [government-funded] booklet has also suggested that
expecting mothers read about the life of great personalities, keep
themselves in "peace" and hang "good and beautiful pictures"
in their bedrooms for a healthy baby. (*Times of India* 2017)

The programs and website make grandiose forecasts, including a
higher IQ, fair skin, and tall stature for the baby and an easy labor for the
mother, as one of the other quotes suggests: "If the mother chants shlokas
and mantras, it helps in the mental growth of the baby . . . if she leads
such a life, there will be no labour pain and the baby will gain up to 300g
more weight" (Ashutosh Bharadwaj 2017; Gowen 2017).

Lest we think these are a few fringe groups, it is important to remember that promotional materials and information are often government
funded and featured on government websites. For example, the government-funded Central Council for Research in Yoga and Naturopathy produced a booklet released by the minister of state for AYUSH that contained
much of this information (*Times of India* 2017). Information to produce
uttam santati (superior children) has made its way into textbooks in some
states. Controversial teachings on how to produce a "superior male child"
through diet and melted gold and silver have found their way into the
curriculum for a Bachelor of Ayurveda, Medicine, and Surgery (a five-
and-a-half-year degree) third-year textbook in the state of Maharashtra
(R. Mishra 2017). While these ideas have long circulated in India, with a
Hindu nationalist government at the helm promoting such knowledge as
Vedic science, these projects are increasingly finding national reach.

The projects and their goals are ambitious. One claims to have already
ensured the delivery of 450 "customised babies," and its target is to have
a Garbh Vigyan Anusandhan Kendra (a facilitation center) in every state
by 2020 (*Indiatimes* 2017). They have also begun to incorporate *garbh vigyan
sanskar* (pregnancy science rites) into college curricula.

Most alarming are the hopeful claims linking their projects to the
successes of Nazi Germany. Several organizers have repeated the narrative that the project was inspired by the advice a senior Hindu nationalist
(RSS) ideologue received over forty years ago in Germany from a woman
he called the "Mother of Germany." The woman is quoted as telling him,

"You have come from India, have you not heard of Abhimanyu (the son of Arjuna in the epic Mahabharata)?" Varshney commented, "She told him that the new generation in Germany was born through Garbh Sanskar and that is why the country is so developed" (Ashutosh Bharadwaj 2017).

POSTCOLONIAL PREGNANCIES:
NEW NARRATIVE POSSIBILITIES

> The abiding western dominology can with religious sanction identify anything dark, profound, or fluid with a revolting chaos, an evil to be mastered, a nothing to be ignored. . . . From the vantage point of the colonizing episteme, the evil is always disorder rather than unjust order; anarchy rather than control, darkness rather than pallor. . . . Yet those who wear the mark of chaos, the skins of darkness, the genders of unspeakable openings—those Others of Order keep finding voice. But they continue to be muted by the bellowing of the dominant discourse.
>
> CATHERINE KELLER, *The Face of the Deep*

This quick tour of the politics of reproduction in postcolonial India highlights the fundamental anxieties of procreation. The political impulses have all centered around eugenic concerns—allowing certain bodies to reproduce while curtailing the reproduction of other bodies. With commercial surrogacy, those with money could hire a gestational surrogate mother to carry their "genetic" child; another person's reproductive future is curtailed through long-term contraception or sterilization. Alongside the growth of gestational surrogacy (now illegal), the sustained focus on population control and family planning continue. The politics of postcolonial India requires understanding these two conditions simultaneously. More importantly, both conditions and their futures are severely circumscribed by their diminished agency. Here, I cannot but contrast these figures with the rich tapestry of Indian mythology. I am also increasingly drawn to the imaginative mythological worlds, albeit not in the supremacist mode that Hindu nationalists invoke, nor for their scientific veracity, but for the liberatory, imaginative, and progressive possibilities imbued in these mythologies. One can read the stories in numerous ways, as science fiction or speculative fiction, as myth, metaphor, or allegory. But they reveal a stunning and pluralistic imagination of the natural and cultural, of gender, sexuality, and kinship. In one possible reading of these worlds, the binary realms of nature and culture, human and nonhuman, and scientific and spiritual mingle to create nonbinary imaginations of naturecultural, human/nonhuman, and spiritual scientific worlds. If we think of these stories alongside what we may call biology, the body, the flesh, or matter, the stories are generative and allow new naturecultural possibilities.

In many mythological stories, nonhumans are bestowed with considerable agencies. Trees communicate with one another and with other inhabitants on earth. Human and nonhuman communication is rampant. Reproduction is not always grounded in sex, heterosexuality, or heteronormative reproduction. Sex and gender are mutable. New bodies can emerge from tissues, through desire, through necessity, through longing, through invocation to the gods. Reproduction is possible in bodies of varying genders, through nonmaterial ghostly worlds, and in magical spectral creatures. Reproduction is possible near and afar, across vast spatial scales. Pregnancy can be entirely unmediated and can occur outside the ontologies of bodies. The metaphoric possibilities of epigenetics are echoed in stories of fetuses in porous wombs, imbibing their cultural surroundings.

What is particularly striking (and what I believe propels Hindu nationalists) is that so much of the mythological imagination that Modi and other Hindu nationalists narrate is now in fact the focus of the cutting edge of contemporary biosciences. Trees do communicate with one another, as do animals; the womb and the fetus do not exist in isolation; reproductive technologies have expanded the possibilities of kinship; and indeed bodies are profoundly mutable. Yet these are not the stories, real or metaphoric, taken up by Hindu nationalists. What is striking about Hindu nationalism or *Hindutva* is their claim of a thoroughly modern, material, global, and scientific Hinduism. They do not rethink or reimagine science; they merely displace a "modern" science from the West and reclaim it within the borders and histories of ancient India, even while they claim the "cultural" modes of a Hinduism steeped in casteist, heterosexist, and patriarchal norms. Unlike other religious fundamentalisms, modern Hinduism has produced not a scriptural fundamentalism but a political nationalism that brings together a melding of science and religion, the ancient and the modern, the past and the present into a powerful brand of nationalism, a vision of India as an archaic modernity.

To understand India is to understand the lush imaginations of the real and unreal, rational and irrational, the natural and preternatural, of this world and others, and learning to see, feel, hear, touch, listen to, and mingle with fellow creatures in this world and others. To embrace the many stolen dreams, fantastical narratives, storytelling traditions, and progressive possibilities that were lost over the long histories of colonialism and conquest. With this embrace, we also open ourselves to the possibilities that in such netherworlds, in their dizzying spirits, in their disavowal of the strictures of rationality, reason, and civilizational logic, we may write new stories for feminism, religion, and science.

◆ AVATAR #6 ◆

The Story of Kalakalappu

Kalakalappu was jubilant. For the first time on Kari, an evolutionary iteration had evolved to embrace its Kanu, or node, of vibrant sociality, mixture, and animism. Of all the Kanus of the Kankavars, this avatar was positioned around the affective and emotional life of the planet. Emotions, the Kankavars believed, were the toughest dimensions of life on Kari. Kalakalappu had no material form—you could only hear it. The other avatars always complained when it sneaked up on them silently. You didn't realize it was around until you heard it! Sometimes it was a gentle giggle, a joyful cheer, or a raucous guffaw that reverberated through the billowy corridors of Avatara Lokam.

Being dull around Kalakalappu was impossible. It was vibrant; it filled minds with ideas, tickled the senses, and energized the body. Kalakalappu made you feel alive—full of vibrancy, joy, and delicious laughter. Through various iterations, the Kankavars had discovered that without inventiveness and creativity, and without maturity and wisdom, emotions quickly devolved into warring factions, into unspeakable violence, leading ultimately to mass killing, genocide, and planetary annihilation. The Kankavars, having learned from their past iterations, waited to introduce advanced emotions late in the planet's evolution; in this latest iteration, they waited to enhance and reinforce the affective and intuitive dimensions of the planet. At last, *Homo impuritus* joined other creatures in a delightful and vibrant planetary polysymphony. Kalakalappu was the final avatar. Once it was unleashed, the Kankavars would deem their experiment successful and would happily leave Kari to its exhilarating futures. After millennia of trying, they believed they were close to finally achieving their goals.

Kalakalappu orchestrated a hubbub of activity. It exulted in its surroundings as sounds reverberated, temporal dimensions collided, spaces imploded, scalar dimensions crisscrossed, visual schemes danced, and an animate planet rejoiced in its newfound cosmic voice. Kalakalappu smiled and took it all in. Through tears in the fabric of time, *Homo impuritus* had learned about the unlimited possibilities of earthly biology. As they gave up distinctions between material and immaterial, the flow of discarded life from Kari to Avatara Lokam became starkly visible. *Homo impuritus* was ashamed. How hostile and unaccepting their planet was! How could they reverse this flow? How might they welcome and engender vital life in everything?

With a singular purpose, *Homo impuritus* valiantly persevered. Slowly over time, the flow to Avatara Lokam reduced, as Kari became increasingly responsive to an astonishing array of life, spirits, and ideas. Through a politics of impurity, *Homo impuritus* learned to talk in cosmic lingos—beyond the strictures of time and space, across the languages of the animate and inanimate, and through the barriers of social and species differences. With their newfound wisdom, they brushed past the competitive struggles of earthly lives to discover new cosmologies of harmony. In these new cosmic configurations, life was no longer reduced to earthly material living; the infinite dimensions of time, space, and matter had opened up endless, exuberant possibilities of living. In the dynamic cosmology of Kalakalappu they saw the pointlessness of factions, violence, and greed. It entertained, educated, lulled, pleasured, calmed, and excited the mind, the senses, and the spirit.

Homo impuritus finally discovered that it was diversity and cooperation, not purity and competition, that were key to wondrous living. They came to embrace the multiplicity and diversity of form, chemistry, space, time, grace, and beauty—*Homo impuritus* had unearthed, understood, and embraced all the clues that the avatars had embedded in earthly cosmology. Rejecting the eternal factionalism of strife with neighbors, this species of humans had opened its hearts and minds and embraced the universe. They had brought together all other life forms across time and space into a creative, imaginative, yet equitable and moral cosmology.

The Kankavars were ecstatic. As they had long believed, carbon-based biology could indeed lead to a vibrant yet peaceful planet, a dynamic yet stable world, an exuberant yet wise people, and an enlightened yet playful cosmology. It is in precisely embracing both sides of the binaries, in rendering oxymoronic syntax into cosmological harmony, that *Homo impuritus* exceeded what had seemed possible in earthly biologies. *Homo impuritus* had come to understand their creators and what they had created, and to make it their own. The creation had now matched their creators, the Kankavars; and like the omniscient Kankavars, *Homo impuritus* emerged successful and spectacular in every way!

Avatars for Dreamers

Narrative's Seductive Embrace

It's a poor sort of memory that only works backwards, says the
White Queen to Alice.

LEWIS CARROLL, *Alice in Wonderland*

At some point, on our way to a new consciousness, we will have to
leave the opposite bank, the split between the two mortal combatants
somehow healed so that we are on both shores at once and, at once,
see through serpent and eagle eyes.

GLORIA E. ANZALDÚA, *Borderlands/La Frontera*

MY FAVORITE STORY IN THE RESEARCH FOR THIS BOOK WAS
undoubtedly the sixteenth-century Telugu writer Pingali Suranna's (2002)
The Sound of the Kiss.[1] In this well-regarded work, the author presents us
with a provocative scenario. Brahma, the Creator of the universe, is asked
by his wife, Saraswati, to tell her a story. Lord Brahma tries. However, each
time he begins a story, he unwittingly brings into existence its characters
and worlds. He is the Creator after all! The novel unfolds as a series of
attempts nested within others. As it turns out, he cannot carry out the
simple task of telling his wife a story without narrating it into creation.
Ultimately, Lord Brahma gives up. He just kisses his wife! This simple
story is a wonderful allegory about the epistemic power of storytellers and
storytelling, and the critical insight that once you are given the power to

tell stories, you are given the power of "worlding," the power to imagine worlds into existence. You decide a world's characters, its plots, its heroes, saviors, and villains.

What a wonderful allegory this is for our times! We live in an era in which the powerful make up stories at will, and their influential allies repeat them, embellish them further, or just stay silent in their complicity. As fictional worlds propagate, the real world seems atwitter! At the same time, the powerful cry foul over news stories, decrying them as "fake news." Other powerful people are outraged, and they call for a regime of truth, facts, and fact checks. Yet others rejoice at the arrival of a purity politics that promises to purge outsiders from their midst and their gates. Through it all, the voices of the powerless are drowned out in a world where increasingly the outrageous competes with the absurd. The point of a public discourse of proliferating stories and fake news is, of course, to obfuscate, to muddy the waters, to dissolve the boundaries between falsity and truth. This is very familiar territory for those in the margins of power—feminists, people of color, the colonized, those from the third world, queer people, those with disabilities—all those "others" who inhabit nonnormative spaces. Those in the margins have always only known a world where their lives can never be acknowledged by power. It is a surreal existence. Those running away from violence, rape, and pillage are them-selves rendered the violent, raping marauders. When those who inhabit lands that were once colonized, that were looted at will by colonizers, demand justice and a return of what was willfully stolen, they are deemed weak, unworthy, and incapable. Those who speak up against the injus-tices of the world are viciously taunted and threatened, murdered in broad daylight, or accused of instigating violence and injustice. Those whose bodies were once property to be bought and sold, and upon whose bodies authorities continue to openly and willfully commit unspeakable violence, are recast as the violent, vicious, and unreasonable people. Those who open up their cities, homes, churches, and temples to others seeking refuge and sanctuary are themselves accused of being illegal criminals who enable violence and lawlessness.

Anyone who has inhabited a space without power understands the workings of power. They well understand that when everyone can make up any story they want, only the powerful will get to tell theirs, and only the powerful will be given the blessings of truth. Feminists have always known this, and known it well.

This book has been for me a lesson in the power of storytelling. I have found the frame of storytelling enlightening and revelatory, making transparent the stakes of knowledge—its production, its circulations, and ultimately its veracity. When power is absolute, when power is concentrated, only the powerful will be the arbiters of truth; only they will be able to tell stories and bring worlds into existence. Across the world we have seen the seductive power of nativist narratives that convince powerful majorities that they are powerless victims. While most Indians die of curable, infectious, and contagious diseases, modern biopolitics spins seductive tales of high-tech precision medicine in genomic futures. Across the world, the idyllic past blends effortlessly with a politics of present injury, to the promise of a brilliant future through a righteous return to an imagined past.

Rather than a paralysis from competing truth claims, if we step back from our surreal world, we can see the intricate networks of the powers of science and religion, and how the volleys of contradictory truth-claims render the circuits of power starkly visible. If there is one insight that emerges through the five cases that animate this book, it is that the act of tracing the arcs of power tells us more about the past, the present, and the future rather than any objective truth or any subjective belief that science or religion can reveal. As Friedrich Nietzsche (2000) evocatively reflected:

> What is the truth? A mobile army of metaphors, metonyms, and anthropomorphisms—in short, a sum of human relations which have been enhanced, transposed, and embellished poetically and rhetorically, and which after long use seem firm, canonical, and obligatory to a people: truths are illusions about which one has forgotten that this is what they are; metaphors which are worn out and without sensuous power; coins which have lost their pictures and now matter only as metal, no longer as coins. [. . .] Yet we still do not know where the drive to truth comes from, for so far we have only heard about the obligation to be truthful which society imposes in order to exist.

It is for this reason that the fields of feminist studies, postcolonial studies, and STS have long been deeply suspicious about "TRUTH"—the absolute, ultimate, invariable, and binding truth; the kind of truth heralded by the high priests of science and the revered leaders of religion. These fields have long known that those in the margins of power will never be the authors or architects, will never have access to the "truth." A quick perusal of science will easily demonstrate how time and time again, and seemingly magically, it is the powerless who are deemed biologically the most deficient. It is the bodies of women that are endlessly policed—as hyperfertile or barren, as surrogates, or as exploited bodies. Similarly, it is the powerless who are relegated to the margins of religion. The impure "untouchables," menstruating bodies, and queer subjects are always left outside the doors of worship. The "pure" are always the powerful, and the "impure," polluting bodies are simultaneously rejected by the two powerful forces of science and religion as inferior and deficient.

Given this history, it is not surprising that feminist studies, postcolonial studies, and STS have been at the forefront of not only critiquing the powerful institutions of science and religion and their perversions of truth, but have also been at the forefront of moving us away from the totalizing language of "objective truth" and "objectivity." These are the fields that understand not only the dangers of truth but also the dangers of RELATIVISM—in both poles, the strictures of power conquer, colonize, and corrupt completely. In their place, feminist STS has suggested epistemologies that embed analyses of power centrally within knowledge practices—such as Donna Haraway's (1988) *situated knowledges* and *partial knowledges*, Sandra Harding's (1991) *strong objectivity* (because unlike scientific objectivity, power is central to its analysis), and Helen Longino's (1990) *science as social knowledge*, or Karen Barad's (2007) *agential realism*. No doubt there are more. In each of these conceptions, the theorists refuse the binaries of science/nonscience, insisting instead on engaged, embodied, and rigorous analyses of the world, grounded in the recognition of power and hierarchies, as they shape the knowledge practices of researchers and readers. These are the spaces that we need to nurture and build.

Feminist studies, postcolonial studies, and STS have also together revealed to us that science and religion are overdetermined categories. Science and religion are not oppositional institutions but are deeply entangled historically. Western science emerged within Christian clerical traditions, and Judeo-Christian imaginaries permeate Western science and medicine. In this book, we have seen how the thigmotropic tendrils of "universal" science, when they travel across the world, always twine into the scaffolding of local topographies to create sciences with local configurations and inflections. Spirit worlds, animist traditions, and indigenous sciences effortlessly mingle with Western science into hybrid and enchanted formations. Monkey gods, sacred traditions, animism, and science mingle effortlessly in a modern India. Indeed, even in the West, rather than enacting a disenchanted, objective enterprise, scientists remain passionate in their engagements with an enchanted nature (Myers 2015; Whitmarsh and Roberts 2016). The vitalist theories of yesteryears are revitalizing the mechanical view of life. The strictly genetic view of reproduction is now being resocialized through epigenetics. At the same time, religions are increasingly embracing science and scientific claims in such areas as evolution, good health, environmentalism, and climate change. We continue to police the boundaries of science and religion as oppositional spaces at our own peril.

Feminists understand power well. If there is one thing that science and religion share, it is that both have been hostile to women and other marginalized groups. Both institutions have historically portrayed women as deficient in their nature and unequal in their capabilities, deemed them unworthy of attention, denied them expert status, and relegated them to the margins of the institution. Feminists understand the importance of the *we*, of working together in collectivity and recognizing that at the heart of purity politics is a purism that is a "de-collectivizing, de-mobilizing, paradoxical, politics of despair" (Shotwell 2016: 9). For these reasons, this particular political moment puts feminists in a significant quandary. In the ongoing "truth wars," one is asked to pick sides, to choose either science or religion; we are asked to march for truth. But the key lessons of feminist studies, postcolonial studies, and STS would suggest that this is no choice for those in the margins. Those in the margins cannot ever win, never mind which side they pick. We must refuse both poles of the binary. Yet we must march for the disenfranchised, the violations on the planet, and the horrors of history, and we must condemn the continued oppressive practices of science and religion and instead embrace their progressive possibilities. The politics of labeling the enemy as "antiscience" is a politics of the powerful seeking to retain their hold on power. Indeed, if we let go of our quest for universal truth, we can avoid the unproductive and pointless politics of antiscience.

We may reject the oppressive histories of science and religion, but as I hope this book has demonstrated, both are deeply social institutions and infinitely malleable to suit varied political ends. Both institutions house a diverse set of practices, histories, and genealogies. Some may find me overly optimistic in my hopes, naive about the likelihood of dismantling the histories of elite power. Perhaps they are right. But my journeys through postcolonial and feminist STS have convinced me that there is no site of purity or innocence—all theories, frameworks, structures, networks, and languages are shaped by violent and oppressive histories. Science and religion are no exceptions. At the same time, I have been deeply moved by how contemporary activists, disenfranchised groups, and some practitioners from within science and religion use their respective institutions toward progressive and feminist goals. We can admire the progressive possibilities of science and religion even as we reject their violent and oppressive histories. Many imaginative possibilities of science and religion are infinitely inviting. How do we live with the earth? How do we live with each other? The interdisciplinary spaces between science and religion, the sites of their entanglements, are fertile sites. They did not accidentally become so. They have been painstakingly nurtured, cultivated, and prepared by generations of feminists and by the many "others" relegated to the margins of history, those who had no access to power or refused power's corruptions. These fertile spaces offer our minds, bodies, and spirits new possibilities, new stories, new imaginations, and the power for new worldings.

I ground this book in stories and the imagination because they, above all, can nurture our speculative futures and our dreams for other worlds. We carry our world, past and future, in our hearts, bodies, spirits, and imaginations, in the stories we tell from the sciences and the humanities. Let us dream, and empower our many feminist avatars to reenchant, revitalize, remember, repopulate, restore, and reenergize our ravaged planet— not to some mythical, nostalgic, pure, or oppressive past but to more imaginative, vibrant, joyous, just, and impure futures.

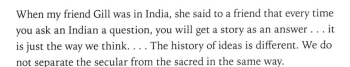

NOTES ON THE MYTHOPOEIA

When my friend Gill was in India, she said to a friend that every time
you ask an Indian a question, you will get a story as an answer . . . it
is just the way we think. . . . The history of ideas is different. We do
not separate the secular from the sacred in the same way.

SUNITI NAMJOSHI AND COOMI VEVAINA,
"An Interview with Suniti Namjoshi"

"We" are the creature that tells itself stories to understand what sort
of creature it is . . . these stories become what we know, what we
understand, and what we are, or, perhaps we should say, what we have
become, or can perhaps be.

SALMAN RUSHDIE, *Two Years Eight Months and Twenty-Eight Nights*

Technoscience, Donna Haraway (cited in Aditya Bharadwaj 2006a) contends, is "a Western Christian salvation" history that promises secular
progress and saves us from apocalypse and damnation. Years later, she
takes this argument further in asking what might have been, had the
technosciences developed within "Buddhist instead of Protestant ways of
worlding" (Haraway 2016: 176). I have been haunted by this question ever
since, wondering about its implications for the multireligious contexts of
Indian cosmologies. The five cases described in this book aim to explore
how the Western Christian salvation narrative plays out in the Indian
context. These illustrative cases show that despite claims of universality,
Western science never unfolds exactly the same everywhere. Rather, it
is always locally inflected. As Indian Nobel laureate and physicist C. V.
Raman once remarked, "I feel it is unnatural and immoral to try to teach
science to children in a foreign language. They will know facts, but they
will miss the spirit" (Wali 2000: 443). As I hope I have revealed, science

in India is varied and multiple, just as there is no singular religion that has shaped India, no one essential way to think or feel Indian. *Indian* is not synonymous with *Europe* or *the West*. These are spaces with different histories, geographies, and cultures, and those differences matter. In science in contemporary India, the Western scripts at times remain eerily similar, except with local modulations. At other times, the narratives are subverted, contradicted, and written anew in the polyphonies of Indian multiplicities and exuberance.

But I also see Haraway's provocative question as an invitation to experiment with genre, especially with fiction and the fantastical. Both unleash narrative resources for other imaginaries and enable thought experiments with the counterfactual. How might the narration of life on Earth unfold in the cadences of Indian storytelling? We hear some of them throughout this book in the claims of Hindu nationalists—claims that Vedic science is a prequel to Western science, completely in harmony with its modern counterpart, or that mythological stories reveal the technoscientific superiority of a precolonial Vedic civilization that colonial rule subsequently eroded. Mythological stories are often offered as evidence of this techno-scientific proficiency:[1] gods traverse the skies in flying chariots (evidence that ancient Hindu civilization had invented flying machines), there are hybrid gods such as Ganesha, a human with an elephant head (evidence that ancient Hindu civilization had advanced medical knowledge to perform plastic surgery), children are born outside the womb (evidence of genetic engineering), Sanjaya could give detailed accounts of battle to King Dhritarashtra in the *Mahabharata* (evidence of ancient internet and satellite technology), and a monkey god and his army of monkeys built a stone bridge across water (evidence that Indian mythology is fact because Rama's Bridge exists today). Mythological stories have been offered to claim pre-Vedic existence for an astonishing array of sciences and discoveries.

It is disconcerting to track how easily Hindu nationalists appropriate South Asian history *as* Hindu history and how uncritically they embrace Western science *as* Vedic science in order to forge the idea of India as a Hindu nation and a global superpower. But I do not want to cede Hinduism to Hindu nationalism. Hinduism is a heterogeneous, diverse, polyvocal, and polytheistic religion that proves to be infinitely flexible and adaptable. There is no singular fundamental text, no singular religious authority, no singular code of conduct, no singular religious ritual, no singular religious practice, no singular god. Hinduism can also be seen as an assemblage of vibrant traditions, including ones that embrace non-violent, pluralistic thinking and an idea of India as a multireligious, secular, democratic republic. While Hinduism and the mythological stories associated with it are laden with histories of caste, class, and gender oppression, these same stories have been embraced and retold in more liberatory renditions. This more capacious reading of India and Hinduism offers tools and imaginaries for more feminist, most just worlds of science and religion. But what is this idea of India?

A. K. Ramanujan (1989), in his well-regarded essay "Is There an Indian Way of Thinking?," heralds the multiplicity of India, but he also answers the title's question in the affirmative. He argues that Indian thinking is marked by many elements, such as eschewing a separation of nature and culture to imagine a "nature-cultural continuum." In contrast to the universality of Western thought, he argues, Indian thought stresses the particular, and always in relational and contextual terms. Therefore, soil produces crops, but it also affects the character of people who live on and feed from the ground. Houses provide shelter, but they also shape the fortunes, mood, and character of people who live in them. Each hour, year, and season has its own properties that portend unique good and bad fortunes for different people. In short, everything depends on context. Musical ragas are attuned to particular times of the day—some robust for the morning, some melodious for unwinding in the evening, others more reflective and sensual for the night. Medicines are personalized depending on the constitution of individuals, never the same for all with the same ailment. Concepts such as *kāma*, *artha*, and dharma are all relational, tied to place, time, personal character, and social role. Contrary to the notion that Indians are spiritual, he insists that they are also materialists, with a strong belief in substance and the flow of substances—"from context to object, from non-self to self" (Ramanujan 1989: 52). Thus, he argues, "all things, even so-called non-material ones like space and time or caste, affect other things because all things are 'substantial' (*dhātu*)" (Ramanujan 1989: 52). And more profoundly, "In Indian medical texts, the body is a meeting-place, a conjunction of elements; they have a physiology, but no anatomy" (Zimmerman, paraphrased by Ramanujan 1989: 52).

In outlining an "Indian" way of thinking, Ramanujan's essay insists on fundamental differences from European and Western thought. In contrast to the facile moves of Hindu nationalists that embrace Western science as Vedic sciences and read mythological fables as historical fact, literary theorists like Ramanujan open a window to a different kind of imagination. They highlight a worldview that is particular, contextual, and relational, that refuses to make easy distinctions between the human and nonhuman, between nature and culture. Such alternate readings draw from a different strand of religions in India: not the supremacist inclinations of India as Hindu, but suffused in India's shared ethos of more liberatory, imaginative, and progressive possibilities imbued in its unique histories, philosophies, and mythologies. In this light, myths that many of us grew up with can be read as astonishingly fertile and radical in their imaginations. To be sure, refusal of binaries does not automatically produce progressive worlds. But it suggests that other worlds are possible.

As I replay my childhood stories, I'm struck by the irony: My postcolonial, British-style education relegated these stories to the realm of silly superstition. Yet having moved to the West, having come to understand how thoroughly colonial imaginations have endured into the postcolonial and the modern, I can only laugh. These mythological stories I grew up with have imaginations that seem au courant, that seem to be what some progressive scholars, indigenous thinkers, environmentalist actors, social justice advocates, anticolonial activists, and contemporary feminists are calling for. I realized that rather than leave behind my childhood stories, I should return to them as generative places to think anew. In many ways, this has been a book about memory—about remembering and remembering and retooling. As science fiction writer Vandana Singh (in Kurtz 2016: 535) remarks, "I am grateful to have been raised on these great epics because they are part of my metaphorical toolbox for making sense of my world. This is likely true for many Indians of my generation."

In invoking the imaginative possibilities of Indian mythology, I don't mean to suggest a utopic tradition. Like Judeo-Christian imaginaries, Hindu mythologies come with their own histories. I also want to heed Anjali Arondekar's (2014: 100) caution against a narrative of simplistic "telos of loss and recovery." Much as I celebrate Ramanujan's recuperation of Indian imagination and thinking, I cannot but approach it with a great deal of carefulness and a great many caveats. I grew up with these stories and have always been deeply aware of how they also perpetuate and naturalize hierarchies of human sociality and organization. These stories, recounted by parents, grandparents, friends, and family and narrated through comic books like the *Amar Chitra Katha* series, were also simultaneously political allegories and morality plays. The stories taught all of us to know our place in society—as members of hierarchies of social organization, central to organized and civil societies. Retold through contemporary lenses, these stories inevitably reified and reproduced social hierarchies and naturalized the political worldviews of their times. As per this logic, of course the lowly rabbit should realize that it was put on Earth as potential prey, and of course the mighty lion was the king of the jungle— every creature had its "rightful place." As Anil Menon (2012: 3) argues, "in all feudal societies, the storytellers were not to disturb the sleep of the privileged few. Predictably, the stories suffered. They could not explore moral, political and social issues with much honesty." Embedded subtly—and sometimes not so subtly—were stories that reinforced the logics of gender, religion, caste, and class: of course the ruling class was always clever and deserving, women were always nurturing and selfless, and the working class best served their purpose through their subjugated labor. Having been schooled in feminist and literary theory, I see now that nothing in these memories remains subtle. A veritable cornucopia of rules of social order were transmitted through these stories, naturalizing the roles of sex, gender, sexuality, caste, and class.

Also important in Ramanujan's essay is what India is not. As Manan Ahmed Asif (2016) points out, "Indian" is mostly synonymous with "Hindu" in Ramanujan's essay, as well as in the works of European philologists like William Jones and Max Müller. However, as he asks:

Is the "Indian" way of thinking evident solely in one grammar, one religion, certain locales, certain specific genres? If there are Buddhist, Jain, and Hindu ways of being Indian, is there no Indian way of being Muslim and no Muslim way of being Indian? If the *Maha-bharata* can elucidate Indian thought, surely that thought can also

belong to a Muslim? . . . We need new histories of our collective
pasts, for we continue to see all pasts through creedal differences.
(Asif 2016: 180)

I find Ramanujan's essay useful because it theorizes distinctly non-
Western narrative possibilities, even though it is not as capacious as I'd
like it to be. It gives us a beginning we can build on. But other sources
help us here. India's significant literary traditions also offer us counter-
mythologies to Brahminical Hinduism.[2] In her wonderful *Seeking Begumpura*, Gail Omvedt (2008) explores the visions of a range of anticaste
intellectuals. Jotirao Phule in his *Gulamgiri* upends the traditional avatar
stories with ones that render the "demon" kings as the strong, valiant,
and righteous and the gods and their avatars as the oppressive settler
colonists (Vendell 2014). The utopia imagined by the Bhakti radical poet-
sant Ravidas (ca. 1450–1520) is a city without sorrow, a casteless, classless
society with no taxes, toil, or hierarchy. In Tukaram's Pandharpur, time
and death have no entry and people dance and mingle with one another;
it is the headman who toils. Kabir's Amrapur is a city of immortality and
Premnagar the city of love (Omvedt 2008). These forceful counternarra-
tives of resistance inspire this mythopoeia.

Given these diverse histories, it is all the more remarkable how Hindu
nationalists erase these shared histories of South Asia to claim a coherent
and unique Indian and Hindu genealogy for the geographies of independent India. It is ironic that the much-celebrated glories of the Vedic period
include areas that are now Pakistan and Afghanistan. The origin stories
of Hindu nationalism constantly elide the tensions of historiography,
creating a unique national prehistory when in fact regional and global
histories are deeply entangled in that prehistory.

South Asia and what is geographically India today is also home to
many religions. Christianity came to India a few centuries after Christ.
Trade relations existed between Arabia and the Indian subcontinent even
in pre-Islamic times. Traders from Judea arrived in India around the sixth
century BCE, and new generations came later as exiles. In short, contemporary India has been a site of economic, political, religious, and spiritual
mingling for millennia. To now erase those many years of coexistence—
albeit through times of war and peace, conquest and rebellion—and now

extract some idea of a "pure" Hinduism is to my mind impossible. Ideas of India today emerge from deeply entangled and intermingled histories. For these reasons, and because of the polytheistic, multireligious, and syncretic traditions of India, I have chosen to use the term *Indian mythology* rather than *Hindu mythology*—because these stories, orally transmitted and circulated over generations, have mingled within the ethos of multireligious and diverse social contexts for millennia.

Interspersed in this book are a series of short fictional interludes. In the mythopoetic Avatara Lokam, the Land of Lost Dreams, these stories imagine new technoscientific avatars. These avatars draw on the stories of the ten avatars (*Dasavatars*) of the god Vishnu. In the original stories, Vishnu descends to Earth during times of crisis to solve problems and restore cosmic equilibrium. For example, in the Narasimha (man-lion) avatar, Vishnu incarnates into a form that is part lion, part human. In the story, a powerful demon-king, Hiranyakashipu, has, through hard work and penance, acquired tremendous powers from the gods. He can never be killed in the daylight or night, in the home or outside, by man or beast. Having checked all the boxes, the demon king feels he is invincible. Confident in these new powers, he creates chaos on Earth, but his son Prahlad, a Vishnu devotee, implores the gods to help avert the chaos. Vishnu creatively adopts the man-lion avatar, arrives at dusk, and kills the demon on the threshold of his palace, thus conforming to yet subverting the demon's powers. In Phule's retelling in *Gulamgiri*, Narasimha kills the demon-king not to protect the gods but to seize his kingdom. Thus, in the mythological and countermythological stories, avatars prove to be exceptional literary devices that effortlessly bridge disparate worlds, allowing us to narrate varied histories, genealogies, and ideologies of oppression, righteousness, and social justice.

The fictional interludes in this book attempt the same. The thigmotropic tales share the book's preoccupation with science and religion, with nature and culture, and with the humanities and the sciences. As feminist scholars have long reminded us, "genres" are themselves epistemological choices of writing. How and what one writes shapes the kinds of insights and understandings that are possible. As I lay out a vision of an "experimental humanities" in the introduction, the fictional interludes take seriously the question of genre and harness the narrative powers of fiction. Experimental fiction as a genre can lay fertile ground for new hypotheses, new questions, new methods. I believe that feminist science studies as a field has inadequately experimented with genre as a form of epistemology. Here, the tradition of experiments in the sciences is valuable methodology for the humanities.

While science fiction in India has older roots (Chattopadhyay 2016), we have recently seen a growth in Indian science fiction writers (V. Singh 2008; Menon and Singh 2012). Although the old mythological tales are riddled with oppressive genealogies, we can still reclaim and reframe these stories to imagine other pasts, presents, and futures. This mythopoeia, drawing on feminist science studies, speculative fiction, and Indian mythology, is one such attempt. Vandana Singh articulates this perfectly:

> Science fiction is the only modern literature I know of where the great questions of our place in the cosmos—things of deep concern to the ancients—are still central. Today we view ourselves as separate from the universe around us, and we see this in mainstream literature where, most of the time, humans only interact with other humans and there is a physical backdrop reality as a kind of static canvas. But the ancients in many cultures were active participants in the cosmos, and they moved comfortably between one reality and another, whether these were physical, metaphoric, or psychological. And the facility that allowed them to do this was the imagination. I consider science fiction and fantasy to be imaginative literature that, at its best, allows us to become participants again. (Kurtz 2016: 545)

Finding India

The Afterlives of Colonialism

आओ कि कोई ख़्वाब बुनें कल के वास्ते
Come, let us weave dreams for tomorrow's sake.

SAHIR LUDHIANVI

Make up a story. . . . For our sake and yours forget your name in
the street; tell us what the world has been to you in the dark places
and in the light. Don't tell us what to believe, what to fear. Show us
belief's wide skirt and the stitch that unravels fear's caul.

TONI MORRISON, Nobel Lecture

THIS BOOK IN MANY WAYS IS AN ODE TO INDIA, AND IN IT, I
have found many Indias. When memories flood me about India, temples
figure prominently. In a book on religious nationalism, it seems fitting to
end at the temple's door. Whatever one's reasons, whether they be religious, spiritual, architectural, or aesthetic, temples in India are incredible
sites to visit. They are sites of the sensual and sensory. Outside the narrow
lanes that lead to most temples is the hustle and bustle of commerce—
selling of religious artifacts and idols as well as offerings to the divine
such as coconuts, fruits, flowers, garlands, incense, and sandalwood. The
temples themselves are often breathtaking; immense structures with
intricate carvings along walls and pillars that are either left as black stone
or painted in vibrant hues against the bright skies. Stories about their

scientific and technological ingenuity accompany any guided tour. Before entering, one is required to remove one's footwear. Immediately, the feet connect to the cold black stone beneath; I remember savoring the cool sensation during many a warm Indian day. The fragrance of the flowers outside lingers on. These smells soon mingle with scents inside—the fragrance of oil, camphor, and the *tirtham*, cool basil-infused holy water that the priest offers to worshippers. I remember running my hands on the cool stone columns as I wandered corridors filled with the echoes of human voices. My fingers would often trace the carvings of figures along the lengths of the temple's exterior and interior walls. I can still hear the rhythmic baritone voice of a priest reciting Vedic chants and Sanskrit *slokas*. The coolness of the temple is complemented by its darkness, the eyes calmed by the shaded corridors away from the intense heat of the sun. These tactile, sensory, and sensual memories, the sights, sounds, and smells, shape my memories of India. India is not only a place; it is also a feeling that overwhelms the senses.

Pico Iyer (2012: 24) in his travel writings argues that for him, "home lies in the things you carry with you everywhere and not the ones that tie you down." As with many immigrants, I too seem to carry home with me wherever I go. Wherever I live, India will always feel a part of me. It is, increasingly, an anachronistic feeling. India has changed, and profoundly. Each time I return, I feel that change in every fiber of my body. The cities I knew, Bombay, Madras, Calcutta, Bangalore, and Trivandrum, are now Mumbai, Chennai, Kolkata, Bengaluru, and Thiruvananthapuram. The ethos of frugality, modesty, and moderation I grew up with have been utterly replaced by conspicuous consumption of mammoth proportions. Relatives visiting me from India look at my life in the United States and inevitably ask, "Same car? Same fridge? Same washing machine?" Yes, I have to say; they all still work!

The conspicuous consumption of the upper middle class is particularly striking because cheaper living is possible, and indeed most Indians live such lives each day. Economic liberalization and globalization have created a robust upper middle class with bottomless pockets. You can eat an elaborate *thali* meal at an expensive restaurant for thousands of rupees or a delicious one next door at a fraction of that cost. A few years ago on my drive from Trivandrum to Nagercoil, the highway was flanked by huge billboards—nearly every one of them for sari and jewelry shops. It was wedding season. The modest yet expensive weddings of yesteryears have been amped up to exorbitant and lavish affairs. Patriarchal norms where only boys' family traditions mattered have given way to a new "feminism" where girls' and boys' family traditions both matter. This has led to a lengthening of wedding rituals and celebrations instead of a neutral or hybrid ceremony, a civil marriage, or a circumvention of marriage altogether. Bollywood film–inspired weddings that feature Punjabi traditions of an evening *sangeet* (singing and dancing alongside abundant libations) are now a fixture even in South India, where most often neither family has even remote connections to Punjab! The wedding reception even in South India has many culinary stations—including North Indian, South Indian, Italian, and Chinese. This rather long description of modern-day India is meant to complement the prologue. My main point is that India still suffuses the senses, but in a different register of class and consumption than just a few decades ago.

In the twenty-first century, global patterns are striking. In the afterlives of colonialisms, and in the midst of the uneven ravages of neocolonial assaults, we have witnessed a surge in the rise of both ethnic nationalism and bionationalism, leading us to a reconstitution of a scientific ethnobionationalism. Much of this book has explored bionationalism in the Indian context. We have seen how genomic technologies have been harnessed to reinvigorate old categories of race and caste into modern subjects and objects through global health and genealogical projects. Old Vedic sciences are given a fresh new patina of the modern as they emerge as marketable consumer products. Women's bodies are commodified anew as they serve as gestational surrogates. Indian ecologies draw on religious mythology to be resanctified. Postcolonial environmentalism has spawned modern gurus and swamis with privatized and for-profit models of worship. The new gurus offer to cure illnesses, cancer, infertility, and homosexuality with the age-old promises of religion. All these developments cohere around a vision of a revitalized Hinduism, one that seeks a modern, scientific, technological, and global reach. These new formations draw clear genealogies from the histories of colonialism, patriarchy, eugenics, and scientific racism into their extensive biopolitical apparatus in the twenty-first century.

Grounded in their nostalgic visions of a "pure and mythical" past, the colorful purity brigades—white, saffron, and green—roam the streets of the world enforcing their collective vision of sameness: ethnic, racial, caste, and religious purity, and purity of nature, knowledge, food, sex and gender norms, culture, and nation. In working on the rise of Hindu nationalism in India, I have come to realize that for all our attempts to provincialize theory, the purity brigades are virulently transnational and diasporic. Yet in tracking the ethno-bionationalism that has emerged in the United States and India, I have found some aspects striking. Those working on the postcolonial world and writing about geographies outside the United States can no longer talk about "special cases" or the "world out there." The metaphoric "third world," the deep inequalities, the "banana republics," the faux democracies, and the plutocracies are now blatantly exposed in the West, just as in the East.

I will admit to a sense of hopelessness about the state of the world today. I am constantly astonished by the bottomless avariciousness, the blatant disregard for suffering, and the utter callousness of politics and policies as they unfold. Yet my search for India leaves me genuinely hopeful and optimistic. In part, my optimism stems from resistance movements across the globe. Recent works in feminism and STS underscore what is by now a fundamental axiom: that power and knowledge are always inextricable. In particular, they highlight how grassroots and subaltern movements across the globe have accessed science to speak to power and to engage in movements toward social justice (Epstein 1996; Philip, Irani, and Dourish 2012; Benjamin 2013; Mamo and Fishman 2013; TallBear 2014; Reardon et al. 2015; Pollock and Subramaniam 2016; STHV 2016; Irani, forthcoming). The five illustrative cases in this book highlight the uneven developments of science in India. Dalit groups access science, just as Hindu nationalist groups do. Science is embraced by many actors alike: the Indian State, grassroots groups, social justice movements, corporations, and the ever proliferating modern religious gurus and their ashrams for transnational devotees. Developing nations like India challenge elite nations in claiming their own genomic sovereignty, and postcolonial environmentalists find unlikely allies in religious nationalists. The landscape is indeed uneven and complex, and many more twists and turns no doubt await.

But at the heart of the unevenness is the increasingly unstable epistemic authority of science. Science, as it was taught to us in our classrooms, is at its heart an enterprise where knowledge is made anew each time new data or information emerges. The spirit of science is an empirical one. Yet through the ravages of conquest, slavery, and colonialism and the continuing horrors of racism, casteism, misogyny, and heterosexism, science continues to assert an absolute authority over "knowledge" in ways that it is unable to sustain. Science is deeply imbricated in these oppressive histories that it has never acknowledged. We cannot write off the dark periods of scientific history to the actions of just a few "bad scientists." A deeper introspection, a more complex analysis is needed. Scientists, and science as an institution, have done a poor job fully understanding and acknowledging their own processes and their location within the networks of power. Within such a history, clarion calls about the importance of scientific truth ring hollow. These calls for truth and facts are tough to sustain when pitted against those who are unable to put food on their table. The rights of an endangered beetle pale against watching your family withering away, with dwindling life expectancies. Claims of the pollution and impurities of food hardly matter when it is the only food that you can afford.

In short, my search for India has taught me the futility of challenging the histories of colonialism and white supremacy with abstract truth-claims and facts. Rather, most compelling are those who can more effectively speak to issues of justice, those who have a path to generating flourishing lives, and those who promise change for impoverished minds and souls. White supremacy and Hindu supremacy are powerful rhetorical tools that have mobilized hope for a future, even when they scapegoat immigrant and minority populations. Religious nationalists have built strong grassroots movements with work on the ground. In elections across the world, it seems clear that we cannot challenge them with abstract notions of absolute truth or facts. Academia, science, and sites of knowledge seem the first places where accusations of elitism stick. What we need in science are more grassroots social movements that translate the possibilities of knowledge into policies and practices that affect the lives on the ground. Instead of isolating ourselves in the name of pure objectivity in the fortress of the ivory tower, we need to move out of the laboratories and into the political landscapes.

The ravages of colonialism tore a world asunder, decimating populations across the globe. Globalization opened up populations for the exploits of capitalism and profit. Now, we have purity brigades who wish to return to enjoy the spoils of a rapacious past by purging the very peoples and the labors that helped produce the wealth. The authoritarian and purity brigades are the true descendants of our colonial histories. For these times, we need theories and politics that work against purity, an impure politics (Shotwell 2016).

I return to my memories of walking down the cool, dark corridors of temples. Today, I see, hear, and feel something different. Alongside the echoes of voices that reverberate through the stone walls, I hear other voices. I hear the lost voices of a more impure past erased from the memories of a more puritanical Hindu present. I do not mean to romanticize this past. It, too, is deeply imbued with power and oppression. I hear the voices of the violence that upper-caste Hinduism has wrought with its many exclusions and excesses. I can see, feel, and hear the technological ingenuity of these massive temple-scapes built in an era before earth-moving machines. I contemplate the lives of the workers who built these temples. How could humans have moved these enormous pieces of rock? Under what conditions, free or enslaved, were these temples built? These memories do give voice to the resplendent figures and stories that emerge from the temple walls—the voluptuous figures that gyrate along the walls, and the many animals and plants that move through the stories. Some of those figures might well be in acts of coercion or in the midst of economic transactions, or they might be individuals with no legal or economic agency. But it reminds us that it wasn't a "golden past," a nostalgic past that a Hindu India must return to.

This is no history of innocence or victimhood. With its strict laws of purity and pollution, these resplendent walls cry out a politics of impurity. I cannot turn my mind away from the fact that most likely the many bodies mobilized to build the temples were the same bodies turned away when in prayer. How many bodies and voices might never have been allowed into these Brahminic hallways through the inhuman dictates of caste and gender that relegated some bodies to exist outside the folds of Hinduism?

Compared to my childhood nostalgia, walking the temple halls today is a much more sobering experience. I am deeply aware of the many inclusions and exclusions of history, some petrified on its walls, others as ghostly presences that wander the halls. To give voice to the marginalized, the excluded, the invisible ghosts that haunt these halls, is to give voice to our impure pasts and our impure present. The heart and strengths of this naturecultural world are always about change, about variation, diversity, and impurity. Justice necessitates a politics of multiplicity, contradiction, ambivalence, and above all resistance. What we need are impure politics, impure theories for impure times.

A NOTE OF GRATITUDE
AND APPRECIATION

It is a pity that academic books and papers are ultimately credited to individual author(s). After all, ideas before they appear in print travel widely and are thoroughly marinated and macerated in circuits of lab group meetings, works-in-progress sessions, conference papers, colloquia, seminars, and many animated conversations with friends and family, and often with strangers. I am so thankful and grateful for the generosity of so many who have indelibly shaped my thinking and understanding over the years. This book introduced me to academic fields that are both new and exciting. It would have been impossible without the faith, support, and kindness of so many. Two individuals in particular were critical for this project. Janet Jakobsen emboldened me to delve into questions of religion, and Sam Hariharan into the unfolding life of India. Without their encouragement and confidence in me, I would not have even dared to undertake this project. Each was instrumental in early formulations of this book project. My deepest gratitude to both.

Two aspects are often critical in the successful completion of projects: first, quiet and generative spaces for work, and second, collegial and generous colleagues. I've been blessed with both over the years. Many thanks to the Suzy Newhouse Center for the Humanities at Wellesley College and its visionary director, Anjali Prabhu. During a fellowship year at the Newhouse Center, I was able to complete this manuscript surrounded by a collegial and supportive environment. Thanks especially to Anjali Prabhu, Claire Fontjn, Sandy Alexandre, Saikat Majumdar, Gurminder Bhogal, Tanalís Padilla, Christopher Polt, Anne Brubaker, Corey McMullen, Peggy Levitt, Jim Petterson, and Smita Radhakrishnan. I am also grateful to the Women's, Gender, and Sexuality Studies Program at

Northeastern University and the Department of the History of Science at Harvard University for making earlier research leaves productive.

This book has been challenging because of its interdisciplinary reach. Many colleagues read drafts of this book and their generosity of time and their critical insights, editorial acumen, intellectual brilliance, and painstaking feedback have enriched this book. Thanks to those who read the entire manuscript. Series editor Rebecca Herzig combined her inimitable characteristics of kindness and openness with an astute and brilliant intellect to provide key advice. Her work has long shaped my thinking, and I'm deeply grateful for her friendship and support. Jyoti Puri's careful and insightful reading considerably strengthened this book. In particular, I am grateful for her help with chapter 2 of this book and for pointing me to key research I had missed, then reading several drafts, each with precision. The mind of Angie Willey, colleague, collaborator, and friend, permeates these pages. Her careful attention, extraordinary interdisciplinary reach, and collaborative spirit have profoundly influenced my work. Jennifer Hamilton has been a constant and extraordinary colleague and friend. Her formidable mind, critical insights, and magnanimous spirit have been so important over the years. Thanks to Manan Ahmed and Durba Mitra for inspiration on the title. A million thanks to Srirupa Roy, Deboleena Roy, Elizabeth Roberts, Sushmita Chatterjee, Durba Mitra, Kalindi Vora, and P K Mahesh, who took valuable time from their busy schedules to read this manuscript. Each proved to be discerning and astute readers, and their keen insights have made this a richer, focused, and more nuanced book than it might have been. I am particularly indebted to the two anonymous reviewers for their careful, generative, and wise reviews that were deeply incisive and insightful. Lengthy conversations with Deboleena Roy and Elizabeth Roberts enriched the discussion of biopolitics. Given the interdisciplinary range of cases in this book, I relied on the wise counsel of many colleagues in developing my analyses. My deepest gratitude for the many conversations over the years, and help with draft chapters: Karen Cardozo, Kiran Asher, Sharmila Rudrappa, Michi Knecht, Ruha Benjamin, Laura Foster, N. Sarojini, Betsy Hartmann, Lisa Armstrong, Linda Blum, Karen Lederer, Sarah Richardson, Rachel Lee, Laura Briggs, Liz Bucar, Marta Calas, Marlene Fried, Anne Pollock, Miliann Kang, Kirsten Leng, Saikat

Majumdar, Itty Abraham, Kavita Philip, Natasha Myers, Priya Kurian, Kentaro Toyama, Karen Lederer, Lynn Morgan, Ahmed Raqab, Sameena Mulla, Vineet Bafna, Brinda Rana, Ramesh Ratnam, Sophia Roosth, Susan Shapiro, Cathy Gere, Ravi Srinivas, Abha Sur, Anil Anantaswamy, Renny Thomas, Kamala Viswesaran, and Suzanna Walters. My gratitude to Avanti Mukherjee for her careful research for chapters 2 and 3.

In this book, I have (with great trepidation!) experimented with speculative fiction as a genre of epistemology. It has been a more intimidating yet exhilarating venture than I expected. I could not have done it without the honest and imaginative feedback of so many. Bodhi Chattopadhyay, Anuj Vaidhya, and Anil Menon in particular have been tremendous interlocutors who have shaped the plot and structure of the fictional interludes. Bodhi's careful and critical reading and suggestions for revisions have left a clear mark on the narrative structure of the book. Many thanks! Enjoyable conversations with Anuj Vaidhya, Xan Chacko, and Ramesh Hariharan helped with the linguistic play with Tamil in the stories. Thanks especially to Ramesh for brainstorming names and characters. Also many thanks to those who engaged with the stories, providing careful and creative feedback and advice on its narrative structure: Janani Balasubramaniam, Betsy Hartmann, Nassim JafariNaimi, Bharat Venkat, Kamakshi Murti, Jehanne Gheith, Ayesha Irani, Daniela Kucz, and Shireen Hamza.

Colleagues in writing groups over the years have also provided fertile grounds for reading and thinking. Thanks to my Boston-based South Asia writing group, Elora Chowdhury Ayeha Irani, Sarah Pinto, and Jyoti Puri; the Five College feminist STS (Science, Technology, and Society) writing group, especially Kiran Asher, Christopher Gunderman, Jennifer Hamilton, Jacquelyne Luce, Lis McLaughlin, Britt Rusert, and Angie Willey; a 4S South Asia STS writing workshop, especially Projit Mukharji, Bharat Venkat, Amit Prasad, Sarah Pinto, and Dwai Banerjee; and the Postcolonial STS Faculty Research Seminar hosted by the Institute for Research on Women and Gender at the University of Michigan: Irina Aristarkhova, Ruha Benjamin, Laura Foster, Sandra Harding, May-Britt Öhman, Anne Pollock, Deboleena Roy, Lindsay Smith, Kim TallBear, Sari Van Anders, and Angie Willey.

The University of Massachusetts Amherst has provided a wonderful home. I am grateful for the collegiality, support, and encouragement of many colleagues over the years. Thanks to members of my department who have enriched my thinking in ways too innumerable to name: Arlene Avakian, Kiran Asher, Laura Briggs, Alex Deschamps, Dayo Gore, Ann Ferguson, Tanisha Ford, Lezlie Frye, Linda Hillenbrand, Karen Lederer, Kirsten Leng, Nancy Patteson, Miliann Kang, Svati Shah, Mecca Jamilah Sullivan, and Angie Willey. I am grateful also for many wonderful colleagues from other departments who have helped with this project: Susan Shapiro, Marta Calas, Stephen Clingman, Laurie Doyle, Asha Nadkarni, Janice Irvine, John Kingston, Joya Misra, and Sangeeta Kamat. In the Five Colleges, I am grateful for the South Asia Studies group, especially Amrita Basu, Lisa Armstrong, Pinky Hota, Krupa Shandilya, Uditi Sen, Priyanka Srivastava, Deepankar Basu, Svati Shah, Nusrat Chowdhary, Shakuntala Ray, Indrani Bhattacharjee, and Dwaipayan Sen.

I could not ask for a more exciting, creative, or brilliant set of colleagues in Feminists in STS (with the indomitable acronym FiSTS) who have over the years made the field such an intellectually vibrant and supportive space—Aimee Bahng, Laura Foster, Carole McCann, Mary Wyer, Sara Giordono, Rajani Bhatia, Angie Willey, Chikako Takeshita, Clare Jen, Jade Sasser, Gwen D'Arcangelis, Harlan Weaver, Cleo Woelfle-Erskine, and Jane Lehr. The field of Feminist STS has been uniquely welcoming to me as an interdisciplinary researcher, providing a fertile and supportive space, nurturing my thinking, and mentoring my professional life. I greatly appreciate the wise counsel and intellectual acuity of so many: Mary Wyer, Angela Ginorio, Evelynn Hammonds, Michelle Murphy, Sharon Traweek, Maralee Mayberry, Karen Barad, Kelly Moore, Rachel Lee, Sandra Harding, Rebecca Herzig, Anne Pollock, Anne Fausto Sterling, Sarah Richardson, Betsy Hartmann, Irina Aristarkhova, Joan Fujimura, Jenny Reardon, Afsaneh Najmabadi, Mel Chen, and Lynn Morgan. Colleagues who work on STS in South Asia are a formidable lot. Conversations and debates have profoundly influenced my scholarship: Geeta Patel, Kath Weston, Amit Prasad, Kavita Philip, Abha Sur, Projit Mukharji, Itty Abraham, Sarah Pinto, Bharat Venkat, Durba Mitra, Kalindi Vora, Lawrence Cohen, N. Sarojini, Shobita Parthasarathy, Priti Ramamurti, Nais Dave, and Smriti Srinivas.

Everyone should be lucky to have an editor like the extraordinary Larin McLaughlin, whose keen intellectual insights, wisdom, encouragement, and firm hand have made this a pleasurable process. Thanks to her, my manuscripts have traveled among the morning glories in Sri Lanka and the beaches of the Atlantic. The incomparable Jennifer Comeau transformed the manuscript with her careful, thorough, and skillful copyediting. My gratitude to the tremendous staff at the University of Washington Press for their hard work in seeing this book through production—Margaret Sullivan, Michael Baccam, and Michael O. Campbell.

My transnational travels and my continuing passion for India have been nurtured through the affections of many loving friends and family members. During my many trips to India, heated discussions and enlivened conversations about everyday life have without doubt left a memorable mark on my experiences of life on the ground. My deep gratitude to the many aunts, uncles, cousins, nieces, and nephews who always welcome me with open arms. Their affection and constant support mean more than I can express. In particular, I always look forward to spirited discussions with Kamal Lodaya, Jayasree, Shari, Jaan, and especially PK Mahesh; Rajan and Malini Sundararajan; Usha, Paapa and P. V. Veeraraghavan; N. Sarojini, and Usha Ramanathan. Thanks to the warm hospitality of Sunderamba and G. N. Mani; Chinni, Vinay, and Sridhar (Padmanabh); Rani and Prasad Subramaniam; Vasanthi and Joy (Parmeshwar); Uthara and Srinath Narayan; Priti Ravi and Rudy Subramaniam; Kanchana and Prakash Hariharan; Raji and Lakshmi Narayan; and Janaki and Rani. A wonderful network of friends and family enlivens my everyday life. I am indebted to my sister Indu Ravi for her unquestioned support for my interests and work schedules. Since I was young, her sharp intellect and independent spirit have been a constant source of inspiration. I am grateful to Kel Moorefield, friend, confidante, and neighbor, for her lively and loving companionship through the years, including many spirited New England evenings. Thanks also to so many friends and family members who make life fun and lively: J, Ravi, Ramesh Ratnam and Asha Ramesh, Arlene Avakian and Martha Ayres, Aditi Subramaniam, Shyam Venugopal, Angie Willey, Jennifer Hamilton, Sada and Anu Warrier, Shreyas Ravi, Daniela Kucz, Lakshmi Ramanathan, Karen Lederer, Brian Sabel, Kiran Asher and Robert Redick, Ani Ravi, Carrie Weng, Betsy Hartmann, Marlene and Bill Fried, Shampa Chanda, V. S. Mani, Kamakshi Murti, Varun Sridhar, Nithya Rathinam, Peggy Schultz, Jim Bever,

Rebecca Dunn, Andrea Kurtz, Rosemary Kalapurakal, Vandana Date, Devamonie Naidoo, Anuradha Goel, Hem Chordia Samdaria, Krupa Shandilya, Nayiree Roubinian, Laurie Doyle, and Jyoti Puri. Thanks to the Zen masters in my life: Leela, who reminds me each day of the joys of the imagination: the pleasures of the tactile and sensory (touch and smell everything!), the rhythm of endlessly and pointlessly throwing stones across the garden, and the thrill of going in search of imaginary mud puddles and then finding them! And the deeply missed Scout, whose loving insistence kept me outdoors during many beautiful New England summer days—wonderful times that helped marinate many ideas in this book. To Sahasranam, the man with a thousand names, a thousand thanks for your sharp intellect, indefatigable curiosity, generous spirit, inimitable humor, and boundless optimism.

If there is one presence I've sorely missed, it is my father, Shankar Subramaniam, a formidable intellect, fierce debater, and lively interlocutor, who died during the initial stages of this project. Throughout this book I have tried to imagine the many spirited discussions and debates we might have had. I can only hope that he would have been proud, and not pronounced me the familial "pound." My mother, Radha Subramaniam, a formidable intellect herself, has shaped my thinking in innumerable ways. She has been a consistent presence and a reliable sounding board, reading much of what I write; I am grateful for her boundless spirit, patience, and unquestioning support.

My eternal gratitude to everyone in my life. Your support has meant more than these meager words convey.

NOTES

PREFACE. IN SEARCH OF INDIA: THE INNER LIVES OF POSTCOLONIALISM

1 Paraphrased from Nehru's famous speech, "Tryst with Destiny," delivered to the Indian Constituent Assembly in the Parliament on August 14, 1947, the eve of India's independence.

2 In 1972, the 42nd Amendment to the Constitution of India changed the description of India from a "sovereign democratic republic" to a "sovereign, socialist, secular democratic republic."

3 The BJP was previously in power as part of a coalition government briefly in 1996 and then again in 1999–2004 (National Democratic Alliance).

INTRODUCTION. AVATARS FOR BIONATIONALISM: TALES FROM (AN)OTHER ENLIGHTENMENT

1 Also important was the series *Chandamama*, and for many of us Indrajal Comics, which featured *The Phantom, Mandrake the Magician, Flash Gordon*, and so on. But none among the latter were locally produced, hence the creation of *Amar Chitra Katha*.

2 Worried that Indian children were getting too Westernized and "alienated from their own culture" (McLain 2009: 24), Anant Pai created the comic series *Amar Chitra Katha* to further "national integration" of India by teaching the "culture, history and ways of life" of India's diverse peoples. The series featured stories from Hindu mythology as well as stories of kings, queens, and other important historical figures in India. Despite these more secular ambitions, studies of *Amar Chitra Katha* argue that Hindu nationalism significantly guided the selection of its heroes, forging a relationship between Hinduism and Indian nationality as well as masculinity, nationalism, and class (Sreenivas 2010).

3 As I explain in the Notes to the Mythopoeia at the end of the book, I am arguably calling these "Indian" mythologies. These mythologies are largely a product of centuries of oral storytelling, drawn from the Hindu pantheon of gods, with regional and transnational variations. However, characters and plotlines from these stories form a metaphoric toolbox for all Indians and permeate much of cultural life in India irrespective of religion (J. Gordon 2016).

4 The Vedic period on the Indian subcontinent occurred roughly between 1700 and 600 BCE, between the end of the urban Indus Valley Civilization and a second urbanization that began in 600 BCE. It is named after the rich and extensive liturgical texts known as the Vedas, the primary sources for our understanding of this period (McClish and Olivelle 2012).

5 The Puranas (literally meaning ancient or old) are a vast collection of writings, composed primarily in Sanskrit. They cover a wide range of topics, including the creation of the universe, religious matters such as customs, ceremonies, sacrifices, caste duties, and so on (*Encyclopedia Britannica* n.d.b).

6 Much of this book is about how "ancient Hindu wisdom" is effortlessly blended with modern science into a seamless whole. However, I should note that there is a small wing of the Hindu right that opposes Western science. For example, India's minister for higher education came under fire for suggesting that Darwin's theory was "scientifically wrong" and needed to change in university curricula (Bagla 2018). Thus far this wing remains at the margins. The minister's comments were followed the next day by the government reaffirming Darwin's theories.

7 Thanks to Pinky Hota for sharing this anecdote.

8 See David Arnold 1993 and 1996.

9 See Abraham 1996, Goonatilake 1984, Nandy 1988, Philip 2004, Rachel Berger 2013b, Mukhaji 2016, Puri 2016, and G. Patel 2017.

10 Many thanks to one of the anonymous reviewers for pushing me to articulate this.

11 Many thanks to one of the anonymous reviewers for this insight.

12 Many thanks to Miliann Kang for suggesting the term *bionarratives*.

CHAPTER 1. HOME AND THE WORLD: THE MODERN LIVES OF THE VEDIC SCIENCES

1 Right after the election, his mother declared that "he will lead the country towards development" (*Hindu* 2014).

2 Modi was chief minister of Gujarat during the 2002 Gujarat riots. There was a systematic targeting of Muslims in mass killings, which many believe were allowed to continue because of the complicity of the state government and law enforcement.

3 Jawaharlal Nehru was India's first prime minister. In addition to secularism, Nehru was deeply committed to science and an industrial model of development.

4 Over the last two decades, religion has become a powerful tool in Indian politics. In fairness to the Hindu nationalists, it must be said that secular governments of the past have also politicized religion (certainly since the time of Indira Gandhi onward) and have also played the "Hindu" card.

5 Patel refers to Sardar Vallabbhai Patel and Subhash refers to Subhash Chandra Bose, both well-known figures in the Indian freedom struggle.

6 Bajrangbali refers to followers of Hanuman, the monkey god in the Hindu pantheon.

7 Although Uma Bharati was given a cabinet position, it was as minister of sports.

8 *Saffronization* has emerged as an Indian political neologism—referring to the saffron robes worn by Hindu *sanyasis* (stage of renunciation or *sanyasa* marking a focus on spiritual life)—to refer to the policies of right-wing Hindu nationalists.

9 Sita and Savitri are both women from Hindu mythology especially renowned for their devotion to their husbands and have come to symbolize the ideals of the "good" Hindu wife.

10 Baba Ramdev was a Tanwar Rajput and saint in the early fifteenth century. Worshipped by Hindus and Muslims, he is believed to have advocated the equality of all human beings. Ramdevra village lies twelve miles from Pokhran.

11 *Yatra* is a spiritual journey usually undertaken as a voyage of discovery or celebration of a significant event, primarily in the role of a witness. However, in its colloquial use, *yatra* simply means "trip" or "journey."

12 It should be noted that the rhetoric of development nationalism is not new. At least since 1987 the patriotism of environmental protesters against "development," especially against dams, has been questioned.

13 Many thanks to one of the anonymous reviewers for this insight.

14 The quotes below contain grammatical and spelling errors—all present in the original. The liberal use of colloquial forms of spoken English in India in these websites is interesting.

15 See, for example, Mallesh Gangaiwar, "Vaastu and Women Health," *Vaastu Doshas and Remedies*, June 23, 2015, www.vaastudoshremedies.com/vaastu-and-women-health/; "Simple Vastu Tips for Better Health," *AstroYatra: An Astrology Portal*, May 30, 2015, https://www.astroyatra.com/blogs/tag/simple-vastu-tips-for-better-health; "Vastu Shastra and Beauty Tips," blog, http://raksha-vastushastrabeautytips.blogspot.com.

16 "Vastu & Woman: Astu and the Women," *Sob Valo*, http://sobvalo.com/vastu-woman/; "Vastu Tips Can Help Women to Deliver Healthy Baby," *Vaastu Shastra: The Encyclopedia of Vastu Shastra*, www.vaastu-shastra.com/women-to-deliver-healthy-baby.html; Nikita Banerjee, "15 Vastu Tips for a Healthy and Smooth Pregnancy," *Astrospeak: Know Your Stars*, February 2, 2017, https://www.astrospeak.com/slides/15-vastu-tips-for-pregnant-women.

17 See, for example, Kasmin Fernandes, "Vastu Ideas for Women," *Times of India*, September 12, 2017, https://timesofindia.indiatimes.com/life-style/home-garden/vaastu-ideas-for-women/articleshow/46501439.cms; "Check Out These Easy Vastu Tips for Women," *Astro Saathi*, April 24, 2017, www.astrosaathi.com/check-out-these-easy-vastu-tips-for-women-1493032477.html; "Vastu Shastra for Court Cases or Settlement Issues or Legal Matters," *Subha Vaastu*, www.subhavaastu.com/court-cases-vastu.html (accessed September 21, 2018); and "Tiny responsible help to our respected woman and Girls in Harassment or bullying issues and prevention of Intended Disturbances through this Vastu Shastra

Website," *Subha Vaastu*, https://www.subhavaastu.com/harrasements-vastu
.html (accessed September 21, 2018).

CHAPTER 2. COLONIAL LEGACIES, POSTCOLONIAL BIOLOGIES: THE QUEER POLITICS OF (UN)NATURAL SEX

Epigraph poem: Suniti Namjoshi and Gillian Hanscombe, *Flesh and Paper* (Prince Edward Island, Canada: Ragweed Press, 1986; Seaton, Devon: Jezebel Tapes and Books, 1986). Heartfelt thanks to Suniti Namjoshi for her kind permission to reproduce these lines.

1 My deepest gratitude to Angie Willey for her engaged reading and deepening my analysis.

2 Many thanks to Deboleena Roy for pushing me to articulate this point.

3 As biology "reenchants" itself, we are returning to these earlier conceptions. From the point of view of some feminist new materialists, perhaps there is a sex life in flowers that is beyond anthropocentrism; it is only the paucity of our frameworks and vocabularies that make it difficult to fully understand and describe the many dimensions of the life of flowers and plants.

4 Many thanks to one of the anonymous reviewers for this insight.

5 A good case in point about the reception of lesbians on screen is the reception of Deepa Mehta's 1996 film *Fire*.

6 See Paragraphs 16.6–16.7 and Paragraph 16.9 in the Supreme Court ruling, "Civil Appeal NO. 10972 OF 2013—Arising out of SLP(C) NO. 15436 of 2009."

7 Many thanks to Sarah Pinto for her critical insight on this point.

8 See Paragraph 38 in the Supreme Court Ruling that details several cases of noncreative and imitative acts of sex for which individuals were booked under Section 377.

9 Paragraph 16.12 in the Supreme Court ruling.

10 Page 38, transcript of the Supreme Court ruling

11 Page 122, transcript of the Supreme Court ruling.

12 Paragraph 16.9 in the Supreme Court ruling.

13 Page 44, transcript of the Supreme Court proceedings.

14 See BJP's statement that decriminalization of homosexuality was the "progressive way" forward (www.livemint.com/Leisure/XCOl7cJw5t3DgnQZsFYIFO/BJP -supports-decriminalization-of-homosexuality-Shaina-NC.html [accessed Sept 9, 2016]) and the RSS spokesperson's comment about re-examining in what circumstances homosexuality could be treated as a crime (www.hindustantimes .com/india/rss-eases-stance-on-decriminalisation-of-gay-sex/story-WCbzY7 ArXmljzzwTZ1ZryO.html).

15 See "Sri Sri Ravi Shankar Speaks on Gay/Homosexuality," December 22, 2012, http://srisriravishankarsrisriravishankar.blogspot.com/2012/12/sri-sri-ravi -shankar-gay.html, and "Homosexuality Not a Crime in India, says Sri Sri

Ravi Shankar," *Firstpost*, December 12, 2013, http://m.firstpost.com/india
/homosexuality-not-a-crime-in-hinduism-says-sri-sri-ravi-shankar-1283843.html
for public statements condemning Section 377.

16 See pages 119–20, transcript of the Supreme Court ruling.

17 On Lord Macaulay's introduction, in 1860, of the concept of "sexual offenses against the order of nature," see Paragraph 64, Delhi High Court Ruling WP(C) No. 7455/2001.

CHAPTER 3. RETURN OF THE NATIVE: NATION, NATURE, AND POSTCOLONIAL ENVIRONMENTALISM

1 See Jaffrelot 2009 and Feagans 2014 for a more detailed history.

2 Deep gratitude to Rebecca Herzig for pushing me to articulate this point.

3 See Gadgil 1985a, 1985b, 1985c, 1987; Gadgil and Vartak 1974; Gadgil and Malhotra 1983; Gadgil and Guha 1992, 1995; and Gadgil et al. 1998.

CHAPTER 4. BIOCITIZENSHIP IN NEOLIBERAL TIMES: ON THE MAKING OF THE "INDIAN" GENOME

1 See National Human Genome Research Institute, "International HapMap Project," https://www.genome.gov/10001688/international-hapmap-project/.

2 "Scheduled castes" and "scheduled tribes" are officially designated groups that have been historically discriminated against. The terms are recognized by the Constitution of India.

3 For more details on the study and a critique of its methods see Subramaniam 2013.

4 This draws from a joint analysis with Projit Mukharji and Shampa Chatterjee. I am indebted to their analyses and insights.

5 Many thanks to Ravi Srinivas for this insight.

6 See PRS Legislative Research, "The National Medical Commission Bill, 2017," www.prsindia.org/billtrack/the-national-medical-commission-bill-2017-5024/.

CHAPTER 5. CONCEIVING A HINDU NATION: (RE)MAKING THE INDIAN WOMB

1 See Akanksha Hospital, "Dr. Nayana H. Patel, Surrogacy and Infertility Expert, Medical Director," http://ivf-surrogate.com/DrNayanaPatel.

2 Many thanks to Kalindi Vora for this insight.

3 See Austa 2014; Bailey 2011, 2014; A. Banerjee 2011; Bhalla and Thapliyal 2013; Dasgupta and Dasgupta 2014; Nayak 2014; Pande 2009a, 2009b, 2010, 2014a, 2014b; Rudrappa 2010, 2015; Sama 2012a, 2012b; and Vora 2013, 2014, 2015a, 2015b.

4 Kalindi Vora, personal communication.

5 Amrita Pande (2014b) argues that some of the "western" narratives of "gift giving," "sisterhood," and "mission" are slowly infiltrating the Indian context, perhaps to minimize the commercial aspect of the enterprise as it gains global attention.

6 There may be differences in the demography of surrogate mothers across different regions in India. See Rudrappa 2015.

CONCLUSION. AVATARS FOR DREAMERS: NARRATIVE'S SEDUCTIVE EMBRACE

1 Many thanks to Anil Menon for introducing me to this story.

NOTES ON THE MYTHOPOEIA

1 There are indeed legitimate claims to a scientific, technological, and philosophically sophisticated civilization: the invention of advanced mathematics, the astronomical calculations of Aryabhata, the medical treatises of Charaka, the linguistic work of Panini, and the settlements of the Indus Valley civilizations (Thatai 2018). However, many of the contemporary claims of the Hindu right I cite do not invoke or draw from this history but instead lie in mythological stories as history.

2 Many thanks to Anuj Vaidya for an introduction to this inspiring literature.

REFERENCES

1KGP (1000 Genomes Project) Consortium. 2015. "A Global Reference for Human Genetic Variation." *Nature* 526, no. 7571: 68–74.

Abdi, S. N. M. 2015. "Saffron Isn't Green: When Religion Is the Pollutant, Can Modi Clean Up Ganga?" *First Post*, May 13, 2015. www.firstpost.com/india/saffron-isnt -green-religion-pollutant-can-modi-clean-ganga-2241984.html.

Abraham, Itty. 1996. "Science and Power in the Postcolonial State." *Alternatives: Global, Local, Political* 21, no. 3 (July–September): 321–39.

———. 1998. *The Making of the Indian Atomic Bomb: Science, Secrecy, and the Postcolonial State*. London and New York: Zed Books.

———. 2006. "The Contradictory Spaces of Postcolonial Techno-Science." *Economic and Political Weekly* 41, no. 3 (January 21–29): 210–17.

———. 2013. "Clashing Cultures of Science." *Postcolonial Studies* 16, no. 4: 406–8.

Abu-Lughod, Lila, ed. 1998. *Remaking Women: Feminism and Modernity in the Middle East*. Princeton, NJ: Princeton University Pres.

Adas, Michael. 1989. *Machines as the Measure of Men: Science, Technology, and Ideologies of Western Dominance*. Ithaca, NY: Cornell University Press.

Agamben, Giorgio. 1995. *Homo Sacer: Sovereign Power and Bare Life*. Stanford, CA: Stanford University Press.

Aggarwal, Shilpi, S. Negi, P. Jha, P. K. Singh, T. Stobdan, M. Q. Pasha, S. Ghosh, A. Agrawal, B. Prasher, M. Mukerji, and S. K. Brahmachari. 2010. "EGLN1 Involvement in High-Altitude Adaptation Revealed through Genetic Analysis of Extreme Constitution Types Defined in Ayurveda." *Proceedings of the National Academy of Sciences* 107, no. 44: 18961–66.

Ahn, S. M., T. H. Kim, S. Lee, D. Kim, H. Ghang, D. S. Kim, B. C. Kim, S. Y. Kim, W. Y. Kim, C. Kim, and D. Park. 2009. "The First Korean Genome Sequence and Analysis: Full Genome Sequencing for a Socio-ethnic Group." *Genome Research* 19, no. 9: 1622–29.

Allender, Tim. 2016. *Learning Femininity in Colonial India, 1820–1932*. Manchester, UK: Manchester University Press.

Alter, Joseph S. 1997. "Seminal Truth: A Modern Science of Male Celibacy in North India." *Medical Anthropology Quarterly* 11: 275–98.

———. 2015. "Nature Cure and Ayurveda: Nationalism, Viscerality and Bio-ecology in India." *Body and Society* 21, no. 1: 3–28.

Alvares, Claude. 1992. *Science, Development, and Violence: The Revolt Against Modernity*. Oxford: Oxford University Press.

Amin, Shaan. 2017. "The Dark Side of the Comics That Redefined Hinduism." *The Atlantic*, December 30, 2017.

Amma. N.d. "GreenFriends." Amma.org. http://amma.org/groups/north-america /projects/green-friends.

Anasuya, Shreya Ila. 2016. "RSS Ambiguity on Homosexuality Masks Larger Problem of Indian Polity." *The Wire*, March 19, 2016.

Andersen, Walter K., and Shridhar D. Damle. 2018. "How the RSS Operates in Foreign Countries in General and the USA in Particular." *Scroll.in*, August 9, 2018. https://scroll.in/article/889509/how-the-rss-operates-in-foreign-countries-in -general-and-the-usa-in-particular.

Anderson, Benedict. 1992. *Long-Distance Nationalism: World Capitalism and the Rise of Identity Politics*. The Wertheim Lecture 1992. Amsterdam: Centre for Asian Studies.

———. 2006. *Imagined Communities: Reflections on the Origin and Spread of Nationalism*. New York: Verso Books.

Anderson, Warwick. 1995. "Excremental Colonialism: Public Health and the Poetics of Pollution." *Critical Inquiry* 21, no. 3: 640–69.

———. 1998. "Where Is the Postcolonial History of Medicine?" *Bulletin of the History of Medicine* 79, no. 3: 522–30.

———. 2002a. *The Cultivation of Whiteness: Science, Health, and Racial Destiny in Australia*. Durham, NC: Duke University Press.

———. 2002b. "Postcolonial Technoscience." *Social Studies of Science* 32 (2002): 643–58.

———. 2007. *Colonial Pathologies: American Tropical Medicine, Race, and Hygiene in the Philippines*. Durham, NC: Duke University Press.

———. 2009. "From Subjugated Knowledge to Conjugated Subjects: Science and Globalization, or Postcolonial Studies of Science? *Postcolonial Studies* 12, no. 4: 389–400.

———. 2017. "Postcolonial Specters of STS." *East Asian Science, Technology and Society: An International Journal* 11, no. 2: 229–33.

Anderson, Warwick, and Vincanne Adams, 2007. "Pramoedya's Chickens: Postcolonial Studies of Technoscience." In *The Handbook of Science and Technology Studies*, edited by Edward J. Hackett, Olga Amsterdamska, Michael Lynch, and Judy Wajcman. Cambridge, MA: MIT Press: 181–203.

Antony, R., and R. Thomas. 2011. "A Mini Review on Medicinal Properties of the Resurrecting Plant *Selaginella bryopteris* (Sanjeevani)." *International Journal of Pharmacy and Life Sciences* 2, no. 7: 933–39.

Anzaldúa, Gloria. 1983. "Speaking in Tongues: A Letter to 3rd World Women Writers." In *This Bridge Called My Back: Writings by Radical Women of Color*, edited by Cherríe Moraga and Gloria Anzaldúa. New York: Kitchen Table: Women of Color Press.

Apffel-Marglin, Frederique, and Pramod Parajuli. 2000. "Sacred Grove and Ecology: Ritual and Science." In *Hinduism and Ecology: The Intersection of Earth, Sky, and*

Water, edited by Christopher Chapple. Cambridge, MA: Harvard University Press: 269–90.

Arnold, David. 1993. *Colonizing the Body: State Medicine and Epidemic Disease in Nineteenth-Century India*. Berkeley: University of California Press.

———. 1996. *Warm Climates and Western Medicine: The Emergence of Tropical Medicine, 1500–1900*. Amsterdam: Rodopi.

———. 2000. "Review of Gyan Prakash (1999) *Another Reason: Science and the Imagination of Modern India*." *Journal of Imperial and Commonwealth History* 28, no 2: 161–63.

———. 2013. "Nehruvian Science and Postcolonial India." *Isis* 104, no. 2: 360–70.

Arondekar, Anjali. 2014. "In the Absence of Reliable Ghosts: Sexuality, Historiography, South Asia." *differences* 25, no. 3: 98–122.

Asad, Talal. 1993. *Genealogies of Religion: Discipline and Reasons of Power in Christianity and Islam*. Baltimore: Johns Hopkins University Press.

Ashok, Sowmiya, and Adrija Roychowdhury. 2018. "The Long Walk: Did the Aryans Migrate into India? New Genetics Study Adds To debate." *Indian Express*, April 16, 2018.

Asif, Manan Ahmed. 2016. *A Book of Conquest*. Cambridge, MA: Harvard University Press.

Atwood, Margaret. 1998. *The Handmaid's Tale*. New York: Anchor Books.

Austa, Sanjay. 2014. "Indian Clinic Brings Hope to Childless Couples and a New Start for Surrogates." *Bangkok Post*, June 23, 2014.

Avanindra, Vaastushastra (pseudonym). 2009. Blog. "What Is Vaastu Shastra?," April 12, 2009, http://vaastushastra-avanindra.blogspot.com/2009/04/what-is-vaastu-shastra.html.

Avestagenome Project. http://theavestagenomeproject.org.

Azam, K. J. 2017. *Rebels from the Mud Houses: Dalits and the Making of the Maoist Revolution in Bihar*. New York: Routledge.

Baber, Zaheer. 1996. *The Science of Empire: Scientific Knowledge, Civilization, and Colonial Rule in India*. Albany: SUNY Press.

Babu, Niranjan. 1998. *Handbook of Vastu*. New Delhi: UBS Publishers' Distributors Ltd.

Bacchetta, Paola. 1999. "When the (Hindu) Nation Exiles Its Queers." *Social Text* 17, no. 4 (61): 141–66.

———. 2013. "Queer Formations in (Hindu) Nationalism." In *Sexuality Studies*, edited by Sanjay Srivastava. New Delhi: Oxford University Press.

Badkhen, Anna. 2015. "Magical Thinking in the Sahel." *New York Times*, June 27, 2015.

Bagchi, Jasodhar. 1990. "Representing Nationalism: Ideology of Motherhood in Colonial Bengal." *Economic and Political Weekly* 25, no. 42/43 (October): 20–27.

———. 1994. *Indian Women: Myth and Reality*, Hyderabad: Sangam Books.

Bagemihl, Bruce. 1999. *Biological Exuberance: Animal Homosexuality and Natural Diversity*. New York: St. Martin's Press.

Bagla, Pallava. 2018. "India's Education Minister Assails Evolutionary Theory, Calls for Curricula Overhaul." *Science*, January 22, 2018. www.sciencemag.org/news

/2018/01/india-s-education-minister-assails-evolutionary-theory-calls-curricula
-overhaul.

Bahiri, Deepika. 2018. *Postcolonial Biology: Psyche and Flesh after Empire*. Minneapolis: University of Minnesota Press.

Bailey, Alison. 2011. "Reconceiving Surrogacy: Toward a Reproductive Justice Account of Indian Surrogacy." *Hypatia* 26, no. 4: 715–41.

———. 2014. "Reconceiving Surrogacy toward a Reproductive Justice Account of Indian Surrogacy." In *Global and Transnational Surrogacy in India: Outsourcing Life*, edited by Sayantini Dasgupta and Shamita Dasgupta. Lanham: Lexington Books: 23–44.

Bakshi, R. 1996. "Development not Destruction: Alternative Politics in the Making." *Economic and Political Weekly* 31, no. 5: 255– 257.

Balasubramanian, D. 2009. "In Search of the Sanjeevani Plant of Ramayana." *The Hindu*, September 10, 2009.

Baldwin, James. 1955. *Notes of a Native Son*. Boston: Beacon Press.

Bamshad, Michael, Alexander Fraley, Michael Crawford, Rebecca Cann, Bharkara Busi, J. M. Naidu, and Lynn B. Jorde. 1996. "mtDNA Variation in Caste Populations of Andhra Pradesh." *Human Biology* 68, no. 1: 1–28.

Bamshad, Michael, T. Kivisild, W. S. Watkins, M. E. Dixon, C. E. Ricker, B. B. Rao, J. M. Naidu, R. B. V. Prasad, P. G. Reddy, A. Rasanayagam, S. S. Papiha, R. Villems, A. J. Redd, M. F. Hammer, S. V. Nguyen, M. Carroll, M. A. Batzer, and L. B. Jorde. 2001. "Genetic Evidence on the Origins of Indian Caste Populations." *Genome Research* 11, no. 6: 994–1004.

Bandyopadhyay, Sibaji. 2007. "Approaching the Present: The Pre-text: The *Fire* Controversy." In *The Phobic and the Erotic: The Politics of Sexualities in Contemporary India*, edited by Brinda Bose and Subhabrata Bhattacharya. Calcutta: Seagull Books: 17–90.

Banerjee, Amrita. 2011. "Reorienting the Ethics of Transnational Surrogacy as a Feminist Paradigm." *The Pluralist* 5: 107–27.

Banerjee, Prathama. 2006. *Politics of Time: "Primitives" and History-Writing in Colonial Society*. Oxford: Oxford University Press.

Banerjee, Sikata. 2003. "Gender and Nationalism: The Masculinization of Hinduism and Female Political Participation in India." *Women's Studies International Forum* 26, no. 2: 167–79.

———. 2006. "Armed Masculinity, Hindu Nationalism, and Female Political Participation in India." *International Feminist Journal of Politics* 8, no. 1: 62–83.

———. 2012. *Make Me a Man! Masculinity, Hinduism, and Nationalism in India*. Albany: SUNY Press.

Barad, Karen. 2007. *Meeting the Universe Halfway: Quantum Physics and the Entanglement of Matter and Meaning*. Durham, NC: Duke University Press.

Baruah, Amit. 2014. "Modi Waxes Eloquent on Women and Goddesses." *The Hindu*, September 2, 2014.

Basalla, George. 1967. "The Spread of Western Science." *Science* 156: 611–22.

Basham, A. L. 1954. *The Wonder That Was India: A Survey of the Culture of the Indian Sub-continent before the Coming of the Muslims*. London: Picador.

Bashford, Alison, and Philippa Levine, eds. 2010. *The Oxford Handbook of the History of Eugenics*. Oxford: Oxford University Press.

Basu, Amrita. 1995. "Introduction." In *Women's Movements in Global Feminism*, edited by Amrita Basu. Boulder, CO: Westview Press: 1–18.

———. 1996. "The Gendered Imagery and Women's Leadership of Hindu Nationalism." *Reproductive Health Matters* 4, no. 8: 70–76.

———. 1998a. "Appropriating Gender." In *Appropriating Gender: Women's Activism and Politicized Religion in South Asia*, edited by Patricia Jeffrey and Amrita Basu. London: Routledge: 3–14.

———. 1998b. "Hindu Women's Activism in India and the Questions It Raises." In *Appropriating Gender: Women's Activism and Politicized Religion in South Asia*, edited by Patricia Jeffrey and Amrita Basu. London: Routledge: 167–84.

Basu, Amrita, and Rekha Basu. 1999. "Of Men, Women, and Bombs: Engendering India's Nuclear Explosions." *Dissent* 46, no. 1 (Winter): 39–43. www.igc.org/dissent.

Basu, Analabha. 2016. "The Dazzling Diversity and the Fundamental Unity: Peopling and the Genetic Structure of Ethnic India." *Indian Journal of History of Science* 51, no. 2.2: 406–16.

Basu, Analbha, Namita Mukherjee, Sangita Roy, Sanghamitra Sengupta, Sanat Banerjee, Madan Chakraborty, Badal Dey, Monami Roy, Bidyut Roy, Nitai P. Bhattacharyya, Susanta Roychoudhury, and Partha P. Majumder. 2003. "Ethnic India: A Genomic View, with Special Reference to Peopling and Structure." *Genome Research* 13, no. 10: 2277–90.

Bauchspies, Wanda, and Maria Puig de la Bellacasa. 2009. "Feminist Science and Technology Studies: A Patchwork of Moving Subjectivities." *Subjectivity* 28: 334–44.

Bauman, Zygmunt. 2013. *Liquid Times: Living in an Age of Uncertainty*. Cambridge, UK: Polity Press.

Baviskar, Amita. 2007. "Indian Indigeneities: Adivasi Engagements with Hindu Nationalism in India." In *Indigenous Experience Today*, edited by Marisol de la Cadena and Orin Starn. New York: Wenner-Gren Foundation for Anthropological Research: 272–303.

———. 2012. "God's Green Earth?" *Himal South Asia*, June 2012. http://m.himalmag .com/gods-green-earth/.

Bayly, Susan. 1995. "Caste and 'Race' in the Colonial Ethnography of India." In *The Concept of Race in South Asia*, edited by Peter Robb. Oxford: Oxford University Press.

———. 2001. *Caste, Society, and Politics in India from the Eighteenth Century to the Modern Age*. Cambridge, UK: Cambridge University Press.

BBC News. 2014. "India's Modi Calls for Reform in Speech to UN." *BBC News*, September 27, 2014. www.bbc.com/news/world-asia-india-29373722.

———. 2017. "What Is India's Caste System?" *BBC News*, July 20, 2017.

Express News Service. 2013. "In Gujarat, Kiran Bedi Heaps Praise on Modi." *Indian Express*, October 23, 2013. http://indianexpress.com/article/cities/ahmedabad/in-gujarat-kiran-bedi-heaps-praise-on-modi/.

Bell, Susan E., and Anne E. Figert, eds. 2015. *Reimagining (Bio)medicalization, Pharmaceuticals, and Genetics: Old Critiques and New Engagements*. New York: Routledge.

Benjamin, Ruha. 2009. "A Lab of Their Own: Genomic Sovereignty as Postcolonial Science Policy." *Policy and Society* 28: 341–55.

———. 2013. *People's Science: Bodies and Rights on the Stem Cell Frontier*. Stanford, CA: Stanford University Press.

——— 2015. "Racial Destiny or Dexterity? The Global Circulation of Genomics as an Empowerment Idiom." In *Reimagining (Bio)medicalization, Pharmaceuticals, and Genetics: Old Critiques and New Engagements*, edited by Susan E. Bell and Anne E. Figert. New York: Routledge: 197–215.

Berger, Rachel. 2013a. "From the Biomoral to the Biopolitical: Ayurveda's Political Histories." *South Asian History and Culture* 4, no. 1: 48–64.

———. 2013b. *Ayurveda Made Modern: Political Histories of Indigenous Medicine in North India, 1900–1955*. Cambridge Imperial and Post-Colonial Studies. New York: Palgrave Macmillan.

Berkes, F., M. Kislalioglu, C. Folke, and M. Gadgil. 1998. "Minireviews: Exploring the Basic Ecological Unit: Ecosystem-Like Concepts in Traditional Societies." *Ecosystems* 1, no. 5: 409–15.

Beteille, Andre. 2001. "Race and Caste." *The Hindu*, March 10, 2001.

Bhabha, Homi, ed. 1990. *Nation and Narration*. New York: Routledge.

Bhalla, Nita, and Mansi Thapliyal. 2013. "India Seeks to Regulate Its Booming 'Rent-a-womb' Industry." Reuters, September 30, 2013. www.reuters.com/article/2013/09/30/us-india-surrogates-idUSBRE98T07F20130930.

Bharadwaj, Aditya. 2006a. "Sacred Modernity: Religion, Infertility, and Technoscientific Conception around the Globe." *Culture, Medicine, and Psychiatry* 30, no. 4 (December): 423–25.

———. 2006b. "Sacred Conceptions: Clinical Theodicies, Uncertain Science, and Technologies of Procreation in India." *Culture, Medicine, and Psychiatry* 30, no. 4 (December): 451–65.

———. 2016. *Conceptions: Infertility and Procreative Technologies in India*. New York: Berghahn Books.

Bharadwaj, Ashutosh, 2017. "RSS Wing Has Prescription for Fair, Tall 'Customised' Babies." *Indian Express*, May 7, 2017.

Bharadwaj, Prajata. 2016. "Where Does the Surrogacy Bill Stand on the Rights of the Surrogate?" *The Wire*, August 31, 2016.

Bhaskaran, Suparna. 2002. "The Politics of Penetration: Section 377 of the Indian Penal Code." In *Queering India: Same-Sex Love and Eroticism in Indian Culture and Society*, edited by Ruth Vanita. New York: Routledge.

———. 2004. *Made in India: Decolonizations, Queer Sexualities, Trans/National Projects*. Basingstoke: Palgrave MacMillan.

Bhatia, Rajani. 2017. *Gender before Birth: Sex Selection in a Transnational Context*. Seattle: University of Washington Press.

Bhosle, Varsha. 1998. "Thirteen at the Table." *Rediff on the Net*, March 19, 1998. www .rediff.com/news/1998/mar/19varsha.htm.

Bhownick, Nilanjana. 2018. "Militant Hinduism and the Reincarnation of Hanuman." *The Wire*, April 4, 2018. https://thewire.in/communalism/noidas-thriving-militant -hinduism-and-the-resurrection-of-hanuman.

Bidwai, Praful. 2007a. "The Question of Faith." *News International*, September 27, 2007.

———. 2007b. "Trumped by a Religious Myth." *News International*, September 22, 2007.

Blackmore, Chelsea. 2011. "How to Queer the Past without Sex: Queer Theory, Feminisms and the Archaeology of Identity." *Archaeologies* 7, no. 1: 75–96.

Bloomberg News. 2017. "Boosting Military Firewpower: Modi Government Close to $660 Million Arms Deal." *Economic Times*, January 4, 2017.

Bode, Maarten. 2015. "Assembling Cyavanaprāsh, Ayurveda's Best-Selling Medicine." *Anthropology and Medicine* 22, no. 1: 23–33.

Bolnick, Deborah A., Duana Fullwiley, Troy Duster, Richard S. Cooper, Joan H. Fujimura, Jonathan Kahn, Jay S. Kaufman, Jonathan Marks, Ann Morning, Alondra Nelson, Pilar Ossorio, Jenny Reardon, Susan M. Reverby, and Kimberly Tall-Bear. 2007. "The Science and Business of Genetic Ancestry Testing." *Science* 318, no. 5849 (October 19): 399–400.

Bose, Brinda, and Subhabrata Bhattacharya. 2007. "Introduction." In *The Phobic and the Erotic: The Politics of Sexualities in Contemporary India*, edited by Brinda Bose and Subhabrata Bhattacharya. Calcutta: Seagull Books: ix–xxxii.

Bose, Purnima. 2008. "Hiindutva Abroad: The California Textbook Controversy." *The Global South* 2, no. 1: 11–34.

Bose, Sugata, and Ayesha Jalal. 2011. *Modern South Asia: History, Culture, Political Economy*. London: Routledge.

Brodwin, Paul. 2002. "Genetics, Identity, and the Anthropology of Essentialism." *Anthropological Quarterly* 75: 323–30.

Brown, Richard P., and Patricia L. Gerbarg. 2005. "Sudarshan Kriya Yogic Breathing in the Treatment of Stress, Anxiety, and Depression, Part I: Neurophysiological Model." *Journal of Alternate and Complementary Medicine* 11, no. 1: 189–201.

Browne, Janet. 1989. "Botany for Gentlemen: Erasmus Darwin and *The Loves of the Plants*." *Isis* 80, no. 4: 592–621.

Burke, Jason. 2014. "Narendra Modi Sworn In as Indian PM in Spectacular Ceremony." *The Guardian*, May 26, 2014. www.theguardian.com/world/2014/may/26 /narendra-modi-sworn-in-india-prime-minister.

Burress, Charles. 2006. "Hindu Groups Lose Fight to Change Textbooks." *SFGate*, March 10, 2006. www.sfgate.com/education/article/SACRAMENTO-Hindu-groups-lose-fight-to-change-2502643.php.

Butalia, Urvashi. 1999. "Soft Target." *Himal South Asian* 12, no. 2 (February). www .himalmag.com.

———. 2000. *The Other Side of Silence: Voices from the Partition of India*. Durham, NC: Duke University Press.

Butalia, Urvashi, and Tanika Sarkar, eds. 1995. *Women and Right-Wing Movements: Indian Experiences*. London: Zed Books.

Caderlof, Gunnel, and K. Sivaramakrishnan, eds. 2006. *Ecological Nationalisms: Nature, Livelihoods, and Identities in South Asia*. Seattle: University of Washington Press.

Campbell, Courtney S. 1992. "Body, Self, and the Property Paradigm." *Hastings Center Report* 22, no. 5: 34–42.

Carney, John. 2015. "Baby Boy Abandoned in India by Parents." *Daily Mail Australia*, June 24, 2015. www.dailymail.co.uk/news/article-3137049/Baby-boy-abandoned -India-surrogate-parents-kept-twin-sister-left-boy-couldn-t-afford-twins.html.

Carney, Scott. 2010. "Inside India's Rent-a-Womb Business." *Mother Jones*, April 8, 2010.

Chakrabarti, Pratik. 2004. *Western Science in Modern India: Metropolitan Methods, Colonial Practices*. Delhi: Permanent Black.

Chakrabarty, Dipesh. 1992. "Postcoloniality and the Artifice of History: Who Speaks for 'Indian' Pasts?" *Representations* 37 (Winter): 1–26.

———. 2000. *Provincializing Europe: Postcolonial Thought and Historical Difference*. Princeton, NJ: Princeton University Press.

———. 2018. "Battle for Minds." *Daily Telegraph*, Tuesday, June 12, 2018.

Chakravarti, Aravinda. 2009. "Human Genetics: Tracing India's Invisible Threads." *Nature*, 461: 487–8.

Chakravarti, Uma. 1993. "Conceptualising Brahmanical Patriarchy in Early India: Gender, Caste, Class, and State." *Economic and Political Weekly* 28, no. 14: 579–85.

Chambers, David Wade, and Richard Gillespie. 2000. "Locality in the History of Science: Colonial Science, Technoscience, and Indigenous Knowledge." In *Nature and Empire: Science and the Colonial Enterprise*, edited by Roy MacLeod. Osiris 15. Chicago: University of Chicago Press: 221–40.

Chambers, John C., James Abbott, Weihua Zhang, Ernest Turro, William R. Scott, et al. 2014. "The South Asian Genome." *PLoS ONE* 9, no. 8: e102645.

Chandra, Vikram. 2014. *Geek Sublime: The Beauty of Code, the Code of Beauty*. Minneapolis: Graywolf Press.

Chatterjee, Meeta. 2000. "Khadi: The Fabric of the Nation in Raja Rao's *Kanthapura*." *New Literatures Review*, no 36: 105–13.

Chatterjee, Partha. 1989. "The Nationalist Resolution of the Women's Question." In *Recasting Women: Essays in Colonial History*, edited by K. Sangari and S. Vaid. Delhi: Kali for Women: 233–53.

———. 1993. *The Nation and Its Fragments: Colonial and Postcolonial Histories*. Princeton, NJ: Princeton University Press.

Chattopadhyay, Bodhisattva. 2016. "On the Mythologerm: Kalpavigyan and the Question of Imperial Science." *Science Fiction Studies* 43, *Indian Science Fiction* (November): 435–558.

Chattopadhyaya, Brajadulal. 1975. *Studying Early India: Archaeology, Texts, and Historical Issues*. New Delhi: Permanent Black.

Chaubey, Gyaneshwer, Anurag Kadian, Saroj Bala, and Vadlamudi Raghavendra Rao. 2015. "Genetic Affinity of the Bhil, Kol and Gond Mentioned in Epic Ramayana." *PloS One* 10, no. 6: p.e0127655.

Chaudhuri, Amit. 2017. "I Am Ramu." *n+1 Magazine*, August 22, 2017. https://nplusonemag.com/online-only/online-only/i-am-ramu/.

Chavda, A. L. 2017. "Aryan Migration Myth." *Indiafacts*, May 5, 2017. http://indiafacts.org/aryan-invasion-myth-21st-century-science-debunks-19th-century-indology/.

Chawla, Raakesh. 1997. *The Pocket Book of Vaastu*. New Delhi: Full Circle.

Chen, Kuan-hsing. 2010. *Asia as Method: Toward Deimperialization*. Durham, NC: Duke University Press.

Chen, Mel. 2012. *Animacies: Bioolitics, Racial Mattering, and Queer Affect*. Durham, NC: Duke University Press.

Chengappa, Raj. 2015. "Making of the Bomb." *India Today*, December 10, 2015.

Chitkara, M. G. 2004. *Rashtriya Swayamsevak Sangh: National Upsurge*. New Delhi: A P H Publishing Corporation.

Chopra, P. 1995. "The South Is Yama's direction." *The Pioneer*, October 20, 1995.

Cipolla, Cyd, Kristina Gupta, David Rubin, and Angela Willey. 2017. *Queer Feminist Science Studies: A Reader*. Seattle: University of Washington Press.

Clarke, Adele, and Donna Haraway, eds. 2018. *Making Kin not Population: Reconceiving Generations*. Chicago: University of Chicago Press.

CNN. 1999. "A Year On, India's Leaders Cheer Its Nuclear Tests." *CNN*, May 11, 1999.

Cohn, Bernard S. 1996. *Colonialism and Its Forms of Knowledge: The British in India*. Princeton, NJ: Princeton University Press.

Cohn, Carol. 1987. "Sex and Death in the Rational World of Defense Intellectuals." *Signs* 12, no. 4: 687–718.

Cooke, Robert. 1999. "Genetic Studies Confirm an Aryan Invasion of India." *Seattle Times*, June 13, 1999.

Coorlawala, Uttara A. 2005. "The Birth of *Bharatanatyam* and the Sanskritized Body." In *Rukmini Devi Arundale, 1904–1986: A Visionary Architect of Indian Culture and the Performing Arts*, edited by Avanthi Meduri. Delhi: Motilal Banarsidass: 173.

Crair, Ben. 2018. "This Multi-billion-Dollar Corporation Is Controlled by a Penniless Yoga Superstar." Bloomberg, March 15, 2018. www.bloomberg.com/news/features/2018-03-15/this-multibillion-dollar-corporation-is-controlled-by-a-penniless-yoga-superstar.

Creanza, N., M. Ruhlen, T. J. Pemberton, N. A. Rosenberg, M. W. Feldman, and S. Ramachandran. 2015. "A Comparison of Worldwide Phonemic and Genetic Variation in Human Populations." *Proceedings of the National Academy of Sciences* 112: 1265–72.

Cronon, William, ed. 1996. *Uncommon Ground: Rethinking the Human Place in Nature*. W. W. Norton and Company.

Crosby, Alfred. 2004. *Ecological Imperialism: The Biological Expansion of Europe, 900–1900*. New York: Cambridge University Press.

Daily Bhaskar. 2013. "Narendra Modi's Gujarat Contributes 40% of Total Surrogacy Market in India." *Daily Bhaskar*, August 11, 2013. http://daily.bhaskar.com/news /GUJ-AHD-narendra-modis-gujarat-contributes-40-of-total-surrogacy-market -in-india-4345295-NOR.html.

Danino, Michel. 2017. "The Problematics of Genetics and the Aryan Issue." *The Hindu*, June 29, 2017.

Daniyal, Shoaib. 2018. "Putting the Horse before the Cart." *Scroll.in*, June 13, 2018. https://scroll.in/article/882188/putting-the-horse-before-the-cart-what-the-dis covery-of-4000-year-old-chariot-in-up-signifies.

Darling, Marsha. 2014. "A Welfare Principle Applied to Children Born and Adopted in surrogacy." Paper presented at the National Women's Studies Association Annual Meeting, Puerto Rico, November 16, 2014.

Das, Pamela, and Richard Horton. 2017. "Pollution, Health, and the Planet: Time for Decisive Action." *The Lancet*, October 19, 2017.

Das, Pinaki. 2007. "Government Should Withdraw ASI's Ram Setu Affidavit and Apologise: Rajnath Singh." *Andhra News*, September 13, 2007. www.andhranews .net/India/2007/September/13-Government-shoulddraw-15466.asp.

Dasa, Gadadhara Pandit. 2012. "The 33 Million Gods of Hinduism." *HuffPost*, August 6, 2012. https://www.huffingtonpost.com/gadadhara-pandit-dasa/the-33-million -demigods-0_b_1737207.html.

Das Acevedo, Deepa. 2013. "Secularism in the Indian Context." *Law and Social Inquiry* 38, no. 1: 138–67.

Dasgupta, Sayantini, and Shamita Dasgupta, eds. 2014. *Global and Transnational Surrogacy in India: Outsourcing Life*. Lanham: Lexington Books.

Dasgupta, Simanti. 2015. *BITS of Belonging Information Technology, Water, and Neoliberal Governance*. Philadelphia: Temple University Press.

Dasgupta, Swapan. 1998. "Indiascope: Issue Date May 25, 1998." *India Today*.

Dave, Naisargi N. 2012. *Queer Activism in India: A Story in the Anthropology of Ethics*. Durham, NC: Duke University Press.

De la Cadena, Marisol. 2010. "Indigenous Cosmopolitics in the Andes: Conceptual Reflections beyond 'Politics.'" *Cultural Anthropology* 25, no. 2: 334–70.

de Mel, Neloufer. 2002. *Women and the Nation's Narrative: Gender and Nationalism in Twentieth Century Sri Lanka*. Lanham, MD: Rowman and Littlefield Publishers.

Deccan Chronicle. 2017. "Hyderabad Police Busts Surrogacy Racket in Hospital." *Deccan Chronicle*, June 18, 2017.

DeLoughrey, Elizabeth, and George B. Handley. 2011. *Postcolonial Ecologies: Literatures of the Environment*. New York: Oxford University Press.

Denyer, Simon. 2011. "India's 'Godmen' Face Questions about Wealth." *Washington Post*, July 12, 2011.

Deomampo, Daisy. 2016. *Transnational Reproduction: Race, Kinship, and Commercial Surrogacy in India*. New York: NYU Press.

Desai, S., and A. Dubey. 2012. "Caste in 21st Century India: COMPETING narratives." *Economic and Political Weekly* 46, no. 11: 40.

Devare, Aparna. 2013. *History and the Making of a Modern Hindu Self.* Abingdon, Oxon: Routledge.

Devi, Maheswata. 1991. "Draupadi." In *Breast Stories*. Calcutta: Seagull Books.

Devrag, Ranjit. 2001. "Study on Caste's Origins in Race Undercuts Govt Stance." Inter Press Service News Agency, August 29, 2001. www.ipsnews.net/2001/08/rights-india-study-on-castes-origins-in-race-undercuts-govt-stance/.

Dhamija, Jasleen. 2000. "Woven Incantations." In *Approaching Textiles, Varying Viewpoints: Proceedings of the Seventh Biennial Symposium of the Textile Society of America*. Santa Fe, NM.

Dhar, Aarti. 2016. "India to Ban Rent-a-Wombs, Limited Surrogacy Allowed but Not for Single Women, Gays." *The Wire*, August 8, 2016.

Dickenson, Donna. 2013. *ME Medicine vs. WE Medicine: Reclaiming Biotechnology for the Common Good*. New York: Columbia University Press.

Dikötter, Frank. 1998. "Race Culture: Recent Perspectives on the History of Eugenics." *American Historical Review* 103, no. 2: 467–78.

Dirks, Nicholas B. 1996. "Recasting Tamil Society: The Politics of Caste and Race in Contemporary Southern India." In *Caste Today*, edited by C. J. Fuller. Delhi: Oxford University Press: 263–95.

———. 2001. *Castes of Mind: Colonialism and Making of Modern India*. Princeton, NJ: Princeton University Press.

Dolnick, Sam. 2011. "Hindus Find a Ganges in Queens, to Park Rangers' Dismay." *New York Times*, April 21, 2011.

Droney, Damien. 2014. "Ironies of Laboratory Work during Ghana's Second Age of Optimism." *Cultural Anthropology* 29, no. 2: 363–84.

D'Souza, Rohan. 2008. "Framing India's Hydraulic Crisis: The Politics of the Modern Large Dam." *Monthly Review* 60, no. 03 (July–August).

Dumit, Joseph. 2012. "Prescription Maximization and the Accumulation of Surplus Health in the Pharmaceutical Industry: The BioMarx Experiment." In *Lively Capital: Biotechnologies, Ethics, and the Governance in Global Markets*, edited by Kaushik Sunder Rajan. Durham, NC: Duke University Press.

Dupré, John. 2002. *Humans and Other Animals*. Oxford: Clarendon Press / Oxford University Press.

Dutt, Barkha. 2018. "India's Judges Make History on Gay Rights—While Its Politicians Fail a Test." *Washington Post*, September 6, 2018.

Economic and Political Weekly. 2005. Editorial. "Setusamudram: Approval without Debate." *Economic and Political Weekly* 40, no. 22–23: 9.

Economic Times. 2014. "Modi's Ram Rajya Remarks No Violation of Law: BJP." *Economic Times*, May 5, 2014.

Economist. 2014. "Landslide for Modi." *The Economist*, May 16, 2014: www.economist.com/blogs/banyan/2014/05/indias-results-roll.

———. 2018. "A New Study Squelches a Treasured Theory about Indians' Origins." *The Economist,* April 5, 2018.

Egorova, Yulia. 2006. *Jews and India: Perceptions and Image.* London: Routledge Curzon.

———. 2010a. "Castes of Genes? Representing Human Genetic Diversity in India." *Genomics, Society, and Policy* 6, no. 3: 32–49.

———. 2010b. "De/geneticizing Caste: Population Genetic Research in South Asia." *Science as Culture* 18, no. 4: 417–34.

———. 2013. "The Substance That Empowers? DNA in South Asia." *Contemporary South Asia* 21, no. 3: 291–303.

Elliott, Carl, and Paul Brodwin. 2002. "Identity and Genetic Ancestry Tracing." *BMJ* 325, no. 7379: 1469–71.

Encyclopedia Britannica. n.d.a. "Adams Bridge." *Encyclopedia Britannica* (online). www .britannica.com/place/Adams-Bridge (accessed November 24, 2017).

———. n.d.b. "India." *Encyclopedia Britannica* (online). www.britannica.com/place /India (accessed January 10, 2018).

Enloe, Cynthia. 1989. *Bananas, Beaches, and Base: Making Feminist Sense of International Politics.* Berkeley: University of California Press.

Epstein, Debbie, Richard Johnson, and Deborah Lynn Steinberg. 2000. "Twice Told Tales: Transformation, Recuperation, and Emergence in the Age of Consent Debates 1998." *Sexualities* 3: 5–30.

Epstein, Steve. 1996. *Impure Science: AIDS, Activism, and the Politics of Knowledge.* Berkeley: University of California Press.

EPW. 2014. Editorial. "Anger, Aspiration, Apprehension." *Economic and Political Weekly* 49, no. 21 (May 24): 7–8.

Escobar, Arturo. 1995. *Encountering Development: The Making and Unmaking of the Third World.* Princeton, NJ: Princeton University Press.

Express Web Desk. 2016. "Cabinet Approves Bill to Prohibit Commercial Surrogacy, Sushma Swaraj Slams Celebrities for Misusing Practice." *Indian Express,* August 25, 2016.

———. 2018. "Supreme Court to Review Section 377." *Indian Express,* January 9, 2018.

Faleiro, Sonia. 2014. "An Attack on Love." *New York Times,* October 31, 2014.

Fan, Chien-Te, Jui-Chu Lin, and Chung-His Lee. 2008. "Taiwan Biobank: A Project Aiming to Aid Taiwan's Transition into a Biomedical Island." *Pharmacogenomics* 9, no. 2: 235–46.

Feagans, Carl T. 2014. "Sacred, Secular, and Ecological Discourses: The Setusamu-dram Project." *Culture and History Digital Journal* 3, no. 1 (June 2014): e009.

Fernandes, Kasmin. 2015. "Vaastu Ideas for Women." *Times of India,* August 3, 2015.

Financial Express. 2018. "What Is National Medical Commission Bill 2017?" *Financial Express,* January 2, 2018. www.financialexpress.com/india-news/what-is-national -medical-commission-bill-2017-the-proposal-that-sparked-pan-india-strike-by -doctors/998567/.

Fischer-Tiné, Harald. 2006. "From Brahmacharya to Conscious Race Culture: Indian Nationalism, Hindu Tradition and Victorian Discourses of Science." In *Beyond Representation. The Construction of Identity in Colonial India*, edited by C. Bates. New Delhi: Oxford University Press: 230–59.

Fish, Allison. 2006. "The Commodification and Exchange of Knowledge in the Case of Transnational Commercial Yoga." *International Journal of Cultural Property* 13, no. 2 (May): 189–206.

Foster, Laura. 2017. *Reinventing Hoodia: Peoples, Plants, and Patents in South Africa*. Seattle: University of Washington Press.

Foucault, Michel. (1976) 1998. *The History of Sexuality*, Volume 1, *The Will to Knowledge*. Translated by Robert Hurley. New edition. New York: Pantheon Books.

———. 1994. *The Birth of the Clinic: An Archaeology of Medical Perception*. New York: Vintage/Random House.

Frenkel, Edward. 2015. "The Reality of Quantum Weirdness." *New York Times*, February 20, 2015.

Fujimura, Joan H. 2000. "Transnational Genomics: Transgressing the Boundary Between the 'Modern/West' and the 'Premodern/East.'" In *Doing Science + Culture*, edited by Roddy Reid and Sharon Traweek. New York: Routledge: 71–92.

Fujimura, Joan, and Ramya Rajagopalan. 2010. "Different Differences: The Use of 'Genetic Ancestry' versus Race in Biomedical Human Genetic Research." *Social Studies of Science* 41, no. 1: 5–30.

Gadgil, Madhav. 1985a. "Towards an Ecological History of India." *Economic and Political Weekly* 20, nos. 45–47: 1909–18.

———. 1985b. "Social Restraints on Resource Utilization: The Indian Experience." In *Culture and Conservation: The Human Dimension in Environmental Planning*, edited by Jeffrey A. McNeely and David C. Pitt. New York: International Union for Conservation of Nature: 135–54.

———. 1985c. "Cultural Evolution of Ecological Prudence." *Landscape and Planning* 12, no. 3: 285–99.

———. 1987. "Diversity: Cultural and Biological." *Trends in Ecology and Evolution* 2, no. 12 (December): 369–73.

Gadgil, Madhav, and Ramachandra Guha. 1992. *This Fissured Land: An Ecological History of India*. Delhi: Oxford University Press.

———. 1995. *Ecology and Equity: The Use and Abuse of Nature in Contemporary India*. New York: Routledge.

Gadgil, Madhav, N. V. Joshi, U. V. Shambu Prada, S. Manoharan, and Suresh Patil. 1998. "Peopling of India." In *The Indian Human Heritage*, edited by D. Balasubramanian and N. Appaji Rao. Hyderabad: Universities Press: 100–129.

Gadgil, Madhav, and Kailash C. Malhotra. 1983. "Adaptive Significance of the Indian Caste System: An Ecological Perspective." *Annals of Human Biology* 10, no. 5: 465–78.

Gadgil, Madhav, and V. D. Vartak. 1974. "Sacred Groves of India: A Plea for Continued Conservation." *Journal of the Bombay Natural History Society* 72, no. 2: 313–26.

Gallais, Pierre, and Vincent Pollina. 1974. "Hexagonal and Spiral Structure in Medieval Narratives." *Yale French Studies*, no. 51, Approaches to Medieval Romance, 115–32.

Ganeri, Jonardon. 2013. "Well-Ordered Science and Indian Epistemic Cultures: Toward a Polycentered History of Science." *Isis* 104: 348–59.

Ganeshaiah, K. N, R. Vasudeva, and R. Uma Shaanker. 2009. "In Search of Sanjeevani." *Current Science* 97, no. 4: 484–89.

Gannett, Lisa. 2001. "Racism and Human Genome Diversity Research: The Ethical Limits of 'Population Thinking.'" *Philosophy of Science* 68, no. 3: S479–S492.

Gettleman, Jeffrey, Kai Schultz, and Suhasini Raj. 2018. "India Gay Sex Ban Is Struck Down. 'Indefensible,' Court Says." *New York Times*, September 6, 2018.

Ghosal, Aniruddha, Kaunain Sheriff, and Pragya Kaushika. "Under 'Ram Avatar Modi', BJP Calls for Ram Rajya in Delhi." *Indian Express*, December 11, 2014.

Ghosh, Amitav. 1988. *The Shadow Lines*. New York: Bloomsbury.

Ghurye, G. S. 1969. *Caste and Race in India*. Bombay: Popular Prakashan.

Gokhale, Namita, and Malashri Lal. 2010. *In Search of Sita: Revisiting Mythology*. New Delhi: Penguin Books.

Gold, Ann Grodzins, and Bhoju Ram Gujar. 1989. "Of Gods, Trees and Boundaries: Divine Conservation in Rajasthan." *Asian Folklore Studies* 48, no. 2: 211–29.

Golwalkar, M. S. 1980. *Bunch of Thoughts*. Bangalore: Jagarana Prakashana.

Goodman, Ellen. 2008. "The Globalization of Baby-Making." *Boston Globe*, April 11, 2008.

Goonatilake, Susantha. 1984. *Aborted Discovery: Science and Creativity in the Third World*. London: Zed Press.

Gopal, Meena. 2017. "Traditional Knowledge and Feminist Dilemmas: Experience of the Midwives of the Barber Caste in South Tamil Nadu." In *Feminists and Science: Critiques and Changing Perspectives in India*, volume 2, edited by Sumi Krishna and Gita Chadha. New Delhi: Sage Press.

Gopalakrishnan, Shankar. 2008. "Neoliberalism and Hindutva." *Countercurrents*, October 30, 2008. www.countercurrents.org/shankar301008.htm.

Gordon, Joan 2016. "Introduction: Indian Science Fiction." "Indian Science Fiction," special issue of *Science Fiction Studies* 43 (November 2016): 433–534.

Gordon, Linda. 2006. "Magic." In *An Introduction to Women's Studies: Gender in a Transnational World*, edited by Inderpal Grewal and Caren Kaplan. Boston: McGraw-Hill.

Goswami, Rakesh. 2017. "From Symbols, History Book Revisions to Cow Minister: BJP-Led Rajasthan Goes Saffron." *Hindustan Times*, April 17, 2017.

Gottweis, Herbert. 2009. Editorial, "Biopolitics in Asia." *New Genetics and Society* 28, no. 3 (September), 201–4.

Gottweis, Herbert, and Byoungsoo Kim. 2009. "Bionationalism, Stem Cells, BSE, and Web 2.0 in South Korea: Toward the Refiguration of Biopolitics." *New Genetics and Society* 28, no. 3: 223–39.

Gowen, Annie. 2017. "Straight Out of the Nazi Playbook." *Washington Post*, May 8, 2017.

Gowen, Annie, and Rama Lakshmi. 2013. "Indian Supreme Court Criminalizes Gay Sex." *Washington Post*, December 11, 2013.

Grosz, Elizabeth. 2011. *Becoming Undone: Darwinian Reflections on Life Politics and Art.* Durham, NC: Duke University Press.

Grove, Richard. 1995. *Green Imperialism: Colonial Expansion, Tropical Island Edens, and the Origins of Environmentalism, 1600–1860.* Cambridge, UK: Cambridge University Press.

Guha, Ramachandra. 2017. "Remembering Nehru in the Time of Modi." *Hindustan Times*, December 17, 2017.

Guha, Sumit. 2013. *Beyond Caste: Identity and Power in South Asia, Past and Present.* Boston: Brill.

Gupta, Dipankar. 2000. *Interrogating Caste: Understanding Hierarchy and Difference in Indian Society.* New Delhi: Penguin Books India.

Gupta, Pooja D. 2015. "Pharmacogenetics, Pharmacogenomics, and Ayurgenomics for Personalized Medicine: A Paradigm Shift." *Indian Journal of Pharmaceutical Sciences* 77, no. 2: 135.

Gurdasani, D., T. Carstensen, F. Tekola-Ayele, L. Pagani, I. Tachmazidou, K. Hatzikotoulas, S. Karthikeyan, L. Iles, M. O. Pollard, A. Choudhury, and G. R. Ritchie. 2015. "The African Genome Variation Project Shapes Medical Genetics in Africa." *Nature* 517 (7534): 327–32.

Haber, M., D. E. Platt, M. A. Bonab, S. C. Youhanna, D. F. Soria-Hernanz, B. Martínez-Cruz, B. Douaihy, M. Ghassibe-Sabbagh, H. Rafatpanah, M. Ghanbari, and J. Whale. 2012. "Afghanistan's Ethnic Groups Share a Y-chromosomal Heritage Structured by Historical Events." *PLOS One* 7, no. 3: p.e34288.

Hall, Stuart. 1996. "When Was the Post-colonial? Thinking at the Limit." In *The Postcolonial Question: Common Skies, Divided Horizons,* edited by Iain Chambers and Lidia Curtis. London: Routledge: 242–60.

Hamilton, Jennifer A. 2015. "Diasporic Proxies and Global Biomedicine." Paper presented at the American Anthropological Association Annual Meeting, Denver, November 19, 2015.

Hamilton, Jennifer A., Banu Subramaniam, and Angela Willey. 2017. "What Indians and Indians Can Teach Us about Colonization: Feminist Science and Technology Studies, Epistemological Imperialism, and the Politics of Difference." *Feminist Studies* 43, no. 3: 612–23.

Hammonds, Evelynn. 1999. "The Logic of Difference: A History of Race in Science and Medicine in the United States." Talk at the Women's Studies Program, UCLA.

Hammonds, Evelynn, and Rebecca M. Herzig. 2009. *The Nature of Difference: Sciences of Race in the United States from Jefferson to Genomics.* Cambridge, MA: MIT Press.

Hannam, James. 2009. *God's Philosophers.* London: Icon Books.

Hansen, Thomas Blom. 1999. *The Saffron Wave: Democracy and Hindu Nationalism in Modern India.* Princeton, NJ: Princeton University Press.

Haraway, Donna J. 1985. *A Manifesto for Cyborgs: Science, Technology, and Socialist Feminism in the 1980s.* San Francisco, CA: Center for Social Research and Education.

———. 1988. "Situated Knowledges: The Science Question in Feminism and the Privilege of Partial Perspective." *Feminist Studies* 14, no. 3 (Autumn): 575–99.

———. 1989. *Primate Visions: Gender, Race, and Nature in the World of Modern Science.* New York: Routledge.

———. 1991. *Simians, Cyborgs, and Women: The Reinvention of Nature.* London and New York: Routledge.

———. 1994. "A Game of Cat's Cradle: Science Studies, Feminist Theory, Cultural Studies." *Configurations* 2, no. 1 (Winter): 59–71.

———. 1997. *Modest_Witness@Second_Millennium.FemaleMan_Meets_OncoMouse: Feminism and Technoscience.* New York: Routledge.

———. 2000. *How Like a Leaf: An Interview with Thyrza Nichols Goodve.* New York: Routledge.

———. 2003. *The Companion Species Manifesto.* Chicago: University of Chicago Press.

———. 2007. *When Species Meet.* Minneapolis: University of Minnesota Press.

———. 2015. "Anthropocene, Capitalocene, Plantationocene, Chthulucene: Making Kin." *Environmental Humanities* 6, no. 1: 159–65.

———. 2016. *Staying with the Trouble: Making Kin in the Chthulucene.* Durham, NC: Duke University Press.

Harding, Sandra. 1991. *Whose Science? Whose Knowledge? Thinking from Women's Lives.* Milton Keynes: Open University Press.

———, ed. 1993. *The "Racial" Economy of Science: Toward a Democratic Future.* Bloomington: Indiana University Press.

———. 1997. "Is Modern Science an Ethno-science? Rethinking Epistemological Assumptions." In *Science and Technology in a Developing World*, edited by T. Shinn, J. Spaapen, and V. V. Krisna. Dordrecht: Springer: 37–64.

———. 1998. *Is Science Multicultural? Postcolonialisms, Feminisms, and Epistemologies.* Bloomington: Indiana University Press.

———. 2008. *Sciences from Below: Feminism, Postcolonialities, and Modernities.* Durham, NC: Duke University Press.

———. 2009. "Postcolonial and Feminist Philosophies of Science And Technology: Convergences and Dissonances." *Postcolonial Studies* 12, no. 4: 401–21.

———. 2012. *The Postcolonial Science and Technology Studies Reader.* Durham, NC: Duke University Press.

Hardy, Billie-Jo, Beatrice Seguin, Peter Singer, Mitali Mukerji, Samir Brahmachari, and Abdallah Daar. 2008. "From Diversity to Delivery: The Case of the Indian Genome Variation Initiative." *Nature Reviews Genetics* 9, S9–S14.

Harris, Gardiner. 2015. "Study Says Pregnant Women in India Are Gravely Underweight." *New York Times*, March 2, 2015.

Harris, Wilson. 1990. "The Fabric of the Imagination." *Third World Quarterly* 12, no. 1: 175–86.

Hartmann, Betsy. 1995. *Reproductive Rights and Wrongs: The Global Politics of Population Control*. Boston: South End Press.

———. 2017. *The America Syndrome: Apocalypse, War, and Our Call to Greatness*, New York: Seven Stories Press.

Hasan, Mushirul. 2002. "Textbooks and Imagined History: The BJP's Intellectual Agenda." *India International Centre Quarterly* 29, no 1: 75–90.

Heath, Deborah., R. Rapp, and K. S. Taussig 2004. "Genetic Citizenship." In *A Companion to the Anthropology of Politics*, edited by David Nugent and Joan Vincent. Malden, MA: Blackwell: 152–67.

Hensman, Rohini. 2014. "The Gujarat Model of Development." *Economic and Political Weekly* 49, no. 11 (March 15).

Herzig, Rebecca. 2005. *Suffering for Science: Reason and Sacrifice in Modern America*. New Brunswick, NJ: Rutgers University Press.

Hindu. 2005. "Sethusamudram Won't Affect Ecology, Says DCI Chief." *The Hindu*, September 26, 2005.

———. 2006. "Setusamudram Project Dredging." *The Hindu*, December 11, 2006. www.thehindubusinessline.com/todays-paper/tp-logistics/Setusamudram-project-dredging/article1754501.ece.

———. 2013a. "Pachauri Warns of Ecological Consequences on Setusamudram." *The Hindu*, April 7, 2013. www.thehindu.com/sci-tech/energy-and-environment/pachauri-warns-of-ecological-consequences-on-Setusamudram/article4591153.ece.

———. 2013b. "Astrology, Vaastu, 100-Feet Long Stage for Modi's UP Rally." *The Hindu*, October 16, 2013. www.thehindu.com/news/national/other-states/astrology-vaastu-100feet-long-stage-for-modis-up-rally/article5240595.ece.

———. 2014. "Modi Will Lead the Country towards Development, Says Mother." *The Hindu*, May 16, 2014.

———. 2017. "65 AYUSH Hospitals in Three Years." October 17, 2017. www.thehindu.com/news/national/time-for-health-revolution-under-the-aegis-of-ayurveda-modi/article19876421.ece.

———. 2018. "Agni-5 Successfully Test-Fired." *The Hindu*, June 4, 2018. www.thehindu.com/todays-paper/tp-national/agni-5-successfully-test-fired/article24074734.ece.

Hindustan Times. 2009. "Poor Women in Gujarat Opt to Become Surrogate Moms." *Hindustan Times*, April 8, 2009.

———. 2014. "PM Narendra Modi Scraps Planning Commission." *Hindustan Times*, August 15, 2014. www.hindustantimes.com/allaboutmodisarkar/pm-modi-announces-scrapping-of-planning-commission-new-body-to-be-set-up-to-foster-cooperative-federalism/article1-1252401.aspx.

———. 2016. "Homosexuality Not a Crime but Socially Immoral: RSS." *Hindustan Times*, March 18, 2016. www.hindustantimes.com/india/homosexuality-shouldnt-be-considered-a-criminal-offence-rss/story-eOZVfy8itwNpH8BZK3l2eM.html.

———. 2017. "Modi Inaugurates All India Institute of Ayurveda." *Hindustan Times*, October 17, 2017. www.hindustantimes.com/india-news/pm-modi-inaugurates-all-india-institute-of-ayurveda-live-updates/story-kcIw85y5ZoHoj3gUiOdBoJ. html.

Hird, Myra J. 2004. *Sex, Gender, and Science*. Basingstoke: Palgrave Macmillan.

Hodges, Sarah. 2008. *Contraception, Colonialism, and Commerce: Birth Control in South India, 1920–1940*. Burlington, VT: Ashgate Publishing.

———. 2010. "South Asia's Eugenic Past." In *The Oxford Handbook of the History of Eugenics*, edited by Alison Bashford and Philippa Levine. Oxford: Oxford University Press.

Hubbard, Ruth. 1990. *The Politics of Women's Biology*. New Brunswick, NJ: Rutgers University Press.

Huggan, Graham, and Helen Tiffin. 2008. "Green Postcolonialism." *Interventions: International Journal of Postcolonial Studies* 9, no. 1, 1–11.

Hustak, C., and N. Myers. 2012. "Involutionary Momentum: Affective Ecologies and the Sciences of Plant/Insect Encounters." *differences* 23, no. 3: 74–118.

IGVC (Indian Genome Variation Consortium). 2005. "Indian Genome Variation Database (IGVdb): A Project Overview." *Human Genetics* 118, no. 1 (November): 1–11.

———. 2008. "Genetic Landscape of the People of India: A Canvas for Disease Gene Exploration." *Journal of Genetics* 87, no. 1: 3–20.

Indiatimes. 2017. "RSS Backed Aarogya Bharati Shares Tips for Women to Deliver." *Indiatimes*, May 7, 2017.

Indian Express. 1998. "VHP Firm on Pokharan Temple, BJP Silent." *Indian Express*, May 25, 1998.

Indo-Asian News Service. 2007. "Advani Calls Up PMO to Protest Ram Sethu Affidavit." *Hindustan Times*, September 13, 2007. www.hindustantimes.com/newdelhi /advani-calls-up-pmo-to-protest-ram-Setu-affidavit/article1-247613.aspx.

Inhorn, Marcia. 2012. "Reproductive Exile in Global Dubai: South Asian Stories." *Cultural Politics* 8: 283–306.

Irani, Lilly. Forthcoming. *Innovators and Their Others: Entrepreneurial Citizenship in Indian Development*. Princeton, NJ: Princeton University Press.

Islam, Md. Nazrul, and K. E. Kuha-Pearce. 2013. "The Promotion of Masculinity and Femininity through Ayurveda in Modern India." *Indian Journal of Gender Studies* 20, no. 3: 415–34.

Iyer, Pico. 2012. *The Man Within My Head*. New York: Knopf.

Jacob, Sharon. 2015. *Reading Mary Alongside Indian Surrogate Mothers: Violent Love, Oppressive Liberation, and Infancy Narratives*. New York: Palgrave Macmillan.

Jaffrelot, Christophe. 2003. *India's Silent Revolution: The Rise of the Lower Castes in North India*. New York: Columbia University Press.

———. 2008. "Hindu Nationalism and the (Not so Easy) Art of Being Outraged: The *Ram Setu* Controversy." *South Asia Multidisciplinary Academic Journal* 2, *Outraged Communities*. https://journals.openedition.org/samaj/1372.

———, ed. 2009. *Hindu Nationalism: A Reader*. Princeton, NJ: Princeton University Press.

Jain, Pankaj. 2011. *Dharma and Ecology of Hindu Communities: Sustenance and Sustainability*. Farnham, Surrey, UK: Ashgate Press.

Jakobsen, Janet, and Ann Pellegrini, eds. 2008. *Secularisms*. Durham, NC: Duke University Press.

Jamal, S., S. Goyal, A. Shanker, and A. Grover. 2017. "Computational Screening and Exploration of Disease-Associated Genes in Alzheimer's Disease." *Journal of Cellular Biochemistry* 118, no. 6: 1471–79.

Jaoul, Nicolas. 2013. "Politicizing Victimhood: The Dalit Panthers' Response to Caste Violence in Uttar Pradesh in the Early 1980's." *South Asian Popular Culture* 11, no. 2: 169–79.

Jasanoff, Sheila. 2011. *Designs on Nature: Science and Democracy in Europe and the United States*. Princeton, NJ: Princeton University Press.

Jasanoff, Sheila, and Sang-Hyun Kim. 2009. "Containing the Atom: Sociotechnical Imaginaries and Nuclear Power in the United States and South Korea." *Minerva* 47, no. 2: 119–46.

Jayaraman, Gayatri. 2013. "The Baby Factory: Surrogacy, the Blooming Business in Gujarat." *India Today*, August 23, 2013.

Jayasekara, Deepa. 2014. "India's Elections: The Decline and Decay of the Congress Party." World Socialist Web Site, April 23, 2004. www.wsws.org/articles/2004/apr2004/indi-a23.sthml.

Jayawardena, K. 1988. *Feminism and Nationalism in the Third World*. London: Zed Books.

Jeffery, Patricia, and Amrita Basu, eds. 1994. *Appropriating Gender: Women's Activism and Politicized Religion in South Asia*. London: Routledge.

Johnston, Josephine, and Mark Thomas. 2003. "Summary: The Science of Genealogy by Genetics." *Developing World Bioethics* 3, no. 2: 103–8.

Jones, Spence. 2014. "How Normal Is Natural? Towards a Deconstruction of Subaltern Sexual Universes and a Redefining of Sexual Activism in the Indian Context." *Critique* (Fall 2014): 89–117.

Joseph, Tony. 2017. "How Genetics Is Settling the Aryan Migration Debate." *The Hindu*, June 16, 2017.

Joshi, P. 1988. *Gandhi on Women*. Delhi: Centre for Development Studies.

Kahn, Jonathan. 2005. "BiDil: False Promises." *Gene Watch* 18, no. 6 (November/December): 6–9.

———. 2013. *Race in a Bottle: The Story of BiDil and Racialized Medicine in a Post-Genomic Age*. New York: Columbia University Press.

Kahn, Susan. 2000. *Reproducing Jews: A Cultural Account of Assisted Conception in Israel*. Durham, NC: Duke University Press.

Kala, Leher. 2014. "Blind Faith." *Indian Express*, November 24, 2014.

Kalra, Bharti, Mahesh Baruah, and Sanjay Kalra. 2016. "The Mahabharata and Reproductive Endocrinology." *Indian Journal of Endocrinology and Metabolism* 20, no. 3: 404–7.

Kannabiran, Kalpana. 2015. "The Complexities of the Genderscape in India." *Seminar* 672 (August): 46–50.

Kapur, Ratna. 2005. *Erotic Justice: Law and the New Politics of Postcolonialism*. Portland, OR: Glasshouse Press.

———. 2007. "The Prurient Postcolonial: The Legal Regulation of Sexual Speech." In *The Phobic and the Erotic: The Politics of Sexualities in Contemporary India*, edited by Brinda Bose and Subhabrata Bhattacharya. Calcutta: Seagull Books: 335–54.

———. 2009. "Out of the Colonial Closet, but Still Thinking 'Inside the Box': Regulating 'Perversion' and the Role of Tolerance in De-radicalising the Rights Claims of Sexual Subalterns." *NUJS Law Review* 2, no. 3: 381–96.

Karlsson, Bengt G. 2006. "Indigenous Natures: Forest and Community Dynamics in Meghalaya, North-East India." In *Ecological Nationalisms: Nature, Livelihoods, and Identities in South Asia*, edited by Gunnel Caderlof and K. Sivaramakrishnan. Seattle: University of Washington Press.

Kathal, P. K. 2005. "Setusamudram Ship Canal Project: Oceanographic/Geological and Ecological Impact on Marine Life in the Gulf of Mannar and Palk Bay, South-eastern Coast of India." *Current Science* 89, no. 7 (October 10): 1082–83.

Katrak, Ketu. 2006. *Politics of the Female Body: Postcolonial Women Writers of the Third World*. New Brunswick, NJ: Rutgers University Press.

Katz, Jonathan. 1990. "The Invention of Heterosexuality." *Socialist Review* 20 (January–March): 7–34.

Keller, Evelyn Fox. 1985. *Reflections on Gender and Science*. New Haven, CT: Yale University Press.

Kelly, Kimberly, and Mark Nichter. 2012. "The Politics of Local Biology in Transnational Drug Testing: Creating (Bio)Identities and Reproducing (Bio)Nationalism through Japanese 'Ethnobridging' Studies." *East Asian Science, Technology, and Society* 6, no. 3: 379–99.

Kent, Eliza F., ed. 2010. "Forests of Belonging: The Contested Meaning of Trees and Forests in Indian Hinduism," special issue of *Journal for the Study of Religion, Nature, and Culture* 4, no. 2.

———. 2013. *Sacred Groves and Local Gods: Religion and Environmentalism in South India*. Oxford: Oxford University Press.

Khalikova, Venera. 2017. "The Ayurveda of Baba Ramdev: Biomoral Consumerism, National Duty and the Politics of 'Homegrown' Medicine in India. *South Asia: Journal of South Asian Studies* 40, no. 1: 105–22.

Kharandikar, Sharvari, Lindsay Gezinski, James Carter, and Marissa Kaloga. 2014. "Economic Necessity or Noble Cause? A Qualitative Study Exploring Motivations for Gestational Surrogacy in Gujarat, India." *Affilia* 29: 224–36.

Kincaid, Jamaica. 1999. *My Garden (Book)*. New York: Macmillan.

Kinsman, Sharon. 2001. "Life, Sex, and Cells." In *Feminist Science Studies: A New Generation*, edited by Maralee Mayberry, Banu Subramaniam, and Lisa Weasel. New York: Routledge: 193–203.

Kishwar, Madhu. 1997. "Yes to Sita, No to Ram." *Manushi* 98: 201–31.

Kivisild, T., M. J. Bamshad, K. Kaldma, M. Metspalu, E. Metspalu, M. Reidla, S. Laos, J. Parik, W. S. Watkins, M. E. Dixon, S. S. Papiha, S. S. Mastana, M. R. Mir, V. Ferak, and R. Villems. 1999. "Deep Common Ancestry of Indian and Western-Eurasian Mitochondrial DNA Lineages." *Current Biology* 9, no. 22: 1331–34.

Kleinman, Arthur. 2016. *Writing at the Margin: Discourses between Anthropology and Medicine.* Berkeley: University of California Press, 1995.

Konduru, Delli Swararao. 2016. "Ethno Medicine and Curative Practices of Savara Tribe in India." *Imperial Journal of Interdisciplinary Research* 2, no. 5: 1079–89.

Kosambi, Damodar Dharmanand. 1988. *The Culture and Civilization of Ancient India in Historical Outline.* Delhi: Vikas.

Kothari, Rajni. 1989. "The Indian Enterprise Today." *Daedalus: Journal of American Academy of Arts and Sciences* (Fall): 51–67.

Kotiswaran, Prabha. 2018. "Surrogacy Regulation Is Stuck between Market, Family, and State." *The Wire*, March 28, 2018. https://thewire.in/women/stuck-between -market-family-and-state-empower-surrogates-themselves.

Koushal and Ors v. Naz Foundation. 2012. Record of the Proceedings in the Supreme Court of India, February 13–March 27, 2012. www.lawyerscollective.org/wp-con tent/uploads/2010/11/Proceedings-of-the-Final-Hearing-in-Section-377-Case.pdf.

Krishna, Sumi, 1996. *Environmental Politics: People's Lives and Development Choices.* New Delhi: Sage Publications.

Krishna, Sumi, and Gita Chadha, eds. 2017. *Feminists and Science: Critiques and Changing Perspectives in India*, Volume 2. New Delhi: Sage Press.

Kumar, Ashutosh, and Deepka Lokhande. 2013. "No Business, Like God Business; India's Godmen Find Spirituality to Be Profitable." *DNA*, September 15, 2013. www.dnaindia.com/lifestyle/report-no-business-like-god-business-indias-godmen -find-spirituality-to-be-profitable-1888934.

Kumar, Megha. 2016. *Communalism and Sexual Violence in India: The Politics of Gender, Ethnicity and Conflict.* London: IB Tauris.

Kumar, Neelam. 2009. *Women and Science in India: A Reader.* Oxford: Oxford University Press.

Kumar, Sanjay. 2014. "India's 'Yoga Ministry' Stirs Doubts among Scientists." *Nature*, November 19, 2014.

Kumudini, N., A. Uma, S. M. Naushad, R. Mridula, R. Borgohain, and V. K. Kutala. 2014. "Association of Seven Functional Polymorphisms of One-Carbon Metabolic Pathway with Total Plasma Homocysteine Levels and Susceptibility to Parkinson's Disease among South Indians." *Neuroscience Letters* 568: 1–5.

Kurien, P. 2015. "Hinduism in the United States." In *Hinduism in the Modern World*, edited by Brian Hatcher. New York: Routledge: 143.

Kurtz, Malisa. 2016. "Alternate Cuts: An Interview with Vandana Singh." *Science Fiction Studies*, 43, *Indian Science Fiction* (November 2016): 534–45.

Lakshmi, Rama. 2016. "India to Propose a Ban on Commercial Surrogacy." *Washington Post*, August 24, 2016.

Lancaster, John. 2011. "Short Cuts." *London Review of Books* 33, no. 4: 24.

Lancaster, Roger N. 2003. *The Trouble with Nature: Sex in Science and Popular Culture.* Berkeley: University of California Press.

Langford, Jean. 2016. "Medical Eschatologies: The Christian Spirit of Hospital Protocol." *Medical Anthropology,* 35 no. 3: 236–46.

Latour, Bruno. 1993. *We Have Never Been Modern.* Cambridge, MA: Harvard University Press.

Law, John, and Wen-yuan Lin. 2017. "Provincializing STS: Postcoloniality, Symmetry, and Method." *East Asian Science, Technology, and Society: An International Journal,* 11, no. 2: 211–27.

Lawyers Collective. 2013. "Supreme Court Reserves Decision on the Recognition of Gender Identity of Transgender Persons in India." Lawyers Collective (Delhi), October 31, 2013. www.lawyerscollective.org/updates/supreme-court-reserves -decision-recognition-gender-identity-transgender-persons-india.

Lee, Suzanne. 2014. "Surrogate Mothers of India's Cradle of the World." Images. Suzanne Lee Photographer. http://suzannelee.photoshelter.com/gallery /Surrogate-Mothers-of-Indias-Cradle-of-the-World/GooooNjTcjUeo9SA/ (accessed August 28, 2014).

Legg, Stephen. 2007. "Beyond the European Province: Foucault and Postcolonialism." In *Space, Knowledge, and Power: Foucault and Geography,* edited by Jeremy W. Cramton and Stuart Elden. Surrey: Ashgate.

Lemke, Thomas. 2011. *Biopolitics: An Advanced Introduction.* New York: NYU Press.

Longino, Helen. 1990. *Science as Social Knowledge: Values and Objectivity in Scientific Inquiry.* Princeton, NJ: Princeton University Press.

Loomba, Ania. 2009. "Race and the Possibilities of Comparative Critique." *New Literary History* 40, no. 3: 501–22.

———. 2015. *Colonialism/Postcolonialism.* London: Routledge.

Luckhurst, Roger. 2006. "Bruno Latour's Scientification: Networks, Assemblages, and Tangled Objects." *Science Fiction Studies* 33, no. 1: 4–17.

MacLeod, Roy. 2000. "Introduction." In *Nature and Empire: Science and the Colonial Enterprise,* edited by Roy MacLeod. Osiris 15. Chicago: University of Chicago Press: 1–13.

Madhukalya, Amrita. 2013. "Rare Unity: Religious Leaders Come Out in Support of Section 377." *DNA,* December 12, 2013. www.dnaindia.com/india/report-rare -unity-religious-leaders-come-out-in-support-of-section-377-1933612.

Mahapatra, Dhananjay. 2008. "Baby Manji's Case Throws Up Need for Law of Surrogacy." *Times of India,* August 25, 2008.

Majumder, Partha. 2010. "The Human Genetic History of South Asia." *Current Biology* 20, no. 4: R184–R187.

Majumder, Partha, and Analbha Basu. 2015. "A Genomic View of the Peopling and Population Structure of India." *Cold Spring Harbor Perspectives in Biology* 7, no. 4: a008540.

Malhotra, Kailash, Yogesh Gokhale, Sudipto Chatterjee, and Sanjiv Srivastava. 2007. *Sacred Groves in India: An Overview.* New Delhi: Bhopal and Aryan Books International.

Malhotra, Rajiv. 2011. "European Misappropriation of Sanskrit Led to the Aryan Race Theory." *Huffington Post*, March 21, 2011. www.huffingtonpost.com/rajiv -malhotra/how-europeans-misappropri_b_837376.html.

Mamo, Laura, and Jennifer Fishman. 2013. "Why Justice? Introduction to the Special Issue on Entanglements of Science, Ethics, and Justice." *Science, Technology, and Human Values* 38, no. 2.

Mandal, S. 2016. "The Wife as an Accomplice: Section 377 and the Regulation of Sodomy in Marriage in India." *Hong Kong Law Journal* 46: 31–46.

Mangharam, Multi Lakhi. 2009. "'Rama, Must I Remind You of Your Divinity?' Locating a Sexualized, Feminist, and Queer Dharma in the *Ramayana*." *diacritics* 39, no. 1: 75–104.

Mani, Lata. 1998. *Contentious Traditions: The Debate on Sati in Colonial India*. Berkeley: University of California Press.

Margulis, Lynn. *Symbiotic Planet: A New Look at Evolution*. New York: Basic Books, 2008.

Margulis, Lynn, and Dorian Sagan. 2003. *Acquiring Genomes: A Theory on the Origin of Species*. New York: Basic Books.

Markens, Susan. 2012. "The Global Reproductive Health Market: U.S. Media Framings and Public Discourses about Transnational Surrogacy." *Social Science and Medicine* 74: 1745–53.

Markowitz, Sally. 2001. "Pelvic Politics: Sexual Dimorphism and Racial Difference." *Signs* 26, no. 2: 389–414.

Mantri, Rajeev. 2014. "Narendra Modi and the Case for Privatisation." *Niti Central*, September 22, 2014. www.niticentral.com/2014/08/26/narendra-modi-and-the -case-for-privatisation-236610.html.

Mawdsley, E. 2006. "Hindu Nationalism, Neo-Traditionalism, and Environmental Discourses in India." *Geoforum* 37, no. 3: 380–90.

McCann, Carole R. 2016. *Figuring the Population Bomb: Gender and Demography in the Mid-Twentieth Century*. Seattle: University of Washington Press,

McClintock, Anne. 1993. "Family Feuds: Gender, Nationalism, and the Family." *Feminist Review* no. 44 (Summer): 61–80.

———. 1995. *Imperial Leather: Race, Gender, and Sexuality in the Colonial Contest*. New York: Routledge.

McClish, Mark, and Patrick Olivelle. 2012. *The Arthasastra: Selections from the Classic Indian Work on Statecraft*. Indianapolis: Hackett.

McConnachie, James. 2008. *The Book of Love: The Story of the Kamasutra*. New York: Macmillan, 2008.

McDonald, I. 1999. "Physiological Patriots? The Politics of Physical Culture and Hindu Nationalism in India." *International Review for the Sociology of Sport* 34, no. 4: 343–58.

McLain, Karline. 2009. *India's Immortal Comic Books: Gods, Kings, and Other Heroes*. Bloomington: Indian University Press.

———. 2011. "The Place of Comics in the Modern Hindu Imagination." *Religion Compass* 5, no. 10: 598–608.

Medina, Jennifer. 2016. "California to Revise How India is Portrayed in Textbooks." *New York Times*, May 19, 2016.

Mehta, Mona. 2010. "A River of No Dissent: Narmada Movement and Coercive Gujarati Nativism." *South Asian History and Culture* 1, no. 4 (October): 509–28.

Meighoo, Sean. 2016. *The End of the West and Other Cautionary Tales*. New York: Columbia University Press.

Menon, Anil. 2012. "Introduction." In *Breaking the Bow: Speculative Fiction Inspired by the Ramayana*, edited by Anil and Vandana Singh. New Delhi: Zubaan Books.

Menon, Anil, and Vandana Singh. 2012. *Breaking the Bow: Speculative Fiction Inspired by the Ramayana*. New Delhi: Zubaan Books.

Menon, Nivedita. 2013a. "The Regulation of Surrogacy in India—Questions and Concerns: SNA." *Kafila*, Jaunary 10, 2013. http://kafila/org/2012/10/the-regulation-of-surrogacy-in-india-question-and-concerns-sama.

———. 2013b. "Ram Setu: The Ecological Argument against the Sethusamudram Project." *Kafila*, February 26, 2013.

Menon, Ritu, and Kamla Bhasin. 1998. *Borders and Boundaries: Women in India's Partition*. New Brunswick, NJ: Rutgers University Press.

Merchant, Carolyn. 1981. *The Death of Nature: Women, Ecology, and Scientific Revolution*. San Francisco: Harper and Row.

Mies, Maria, and Vandana Shiva. 1993. *Ecofeminism*. Halifax: Fernwood Publishers.

Mishra, Pankaj. "India at 70, and the Passing of Another Illusion." *New York Times*, August 11, 2017.

Mishra, Rashmi. 2017. "How to Conceive a Baby Boy?" *India.com*, May 8, 2017. www.india.com/buzz/how-to-conceive-a-baby-boy-how-to-be-intimate-to-deliver-a-stronger-taller-fairer-and-smarter-kid-where-is-india-going-2109534/.

Momin, Sajeda. 2008. "World Meet to Save Ram Setu." DNA, November 11, 2008.

Moorjani, P., K. Thangaraj, N. Patterson, M. Lipson, P. R. Loh, P. Govindaraj, B. Berger, D. Reich, and L. Singh. 2003. "Genetic Evidence for Recent Population Mixture in India." *American Journal of Human Genetics* 93: 422–38.

Moreland, J. P. 1989. *Christianity and the Nature of Science*. Grand Rapids, MI: Baker Books.

Morris, Andrew, and Robert N. Lightowlers. 2000. "Can Paternal mtDNA Be Inherited?" *The Lancet* 355 (April 15): 1290–91.

Mountain, Joanna L., Joan M. Hebert, Silanjan Bhattacharyya, Peter A. Underhill, Chris Ottolenghi, Madav Gadgil, and L. Luca Cavalli-Sforza. 1995. "Demographic History of India and mtDNA-Sequence Diversity." *American Journal of Human Genetics* 56: 979–92.

Mukharji, Projit B. 2014. "Vishalyakarani as Eupatorium ayapana: Retro-botanizing, Embedded Traditions, and Multiple Historicities of Plants in Colonial Bengal, 1890–1940." *Journal of Asian Studies* 73, no. 1: 65–87.

———. 2016. *Doctoring Traditions: Ayurveda, Small Technologies, and Braided Sciences*. Chicago: University of Chicago Press.

Mukunth, Vasudevan. 2015. "An Upvote for Ayurveda from the Swiss Government—Alongside Homeopathy." *The Wire*, May 15, 2015.

Murphy, Michelle. 2017. *Economization of Life: Calculative Infrastructures of Population and Economy*. Durham, NC: Duke University Press.

Museka, G., and M. M. Madondo. 2012. "The Quest for a Relevant Environmental Pedagogy in the African Context: Insights from Unhu/Ubuntu Philosophy." *Journal of Ecology and the Natural Environment* 4, no. 10: 258–65.

Mutua, Kagendo, and Beth Blue Swadener. 2004. *Decolonizing Research in Cross-Cultural Contexts: Critical Personal Narratives*. Albany: SUNY Press.

Myers, Natasha. 2015. Rendering *Life Molecular: Models, Modelers, and Excitable Matter*. Durham, NC: Duke University Press.

Nadimpally, Sarojini, Deepa Venkatachalam, and Aswathy Raveendran. 2017. "Panel Report on Surrogacy Bill Is a Good Step, but Definitely Isn't Perfect." *The Wire*, August 16, 2017.

Nagar, Malti, and S. C. Nanda. 1986. "Ethnographic Evidence for the Location of Ravana's Lanka." *Bulletin of the Deccan College Research Institute* 45: 71–77.

Nair, Avinash. 2012. "Gujarat Government's Magazine Bats for Surrogacy." *Indian Express*, May 21, 2012.

Nair, Janaki, and Mary E. John, eds. 2000. *A Question of Silence? The Sexual Economies of Modern India*. New York: Zed Books.

Najar, Nida, and Suhasini Raj. 2015. "Desperate Families in Delhi as Dengue Overwhelms Hospitals." *New York Times*, October 9, 2015.

Nakahashi, Wataru, Joe Yuichiro Wakano, and Joseph Henrich. 2012. "Adaptive Social Learning Strategies in Temporally and Spatially Varying Environments." *Human Nature* 23, no. 4: 386–418.

Namjoshi, Suniti, and Coomi Vevaina. 1998. "An Interview with Suniti Namjoshi." *ARIEL: A Review of International English Literature* 29, no. 1 (January): 195–201. https://journalhosting.ucalgary.ca/index.php/ariel/article/viewFile/33940/27979.

Namjoshi, Suniti, and Gillian Hanscombe. 1986. *Flesh and Paper*. Prince Edward Island, Canada: Ragweed Press, 1986; Seaton, Devon: Jezebel Tapes and Books.

Nanda, Meera. 1997. "The Science Wars in India." *Dissent* 44 (Winter), at www. igc .org/dissent.

———. 1998. "Reclaiming Modern Science for Third World Progressive Social Movements." *Economic and Political Weekly* 33, no. 16: 915–22.

———. 2001. "We Are All Hybrids Now: The Dangerous Epistemology of Postcolonial Populism." *Journal of Peasant Studies* 28, no. 2: 162–86.

———. 2003. *Prophets Facing Backward: Postmodern Critiques of Science and Hindu Nationalism in India*. New Brunswick, NJ: Rutgers University Press.

———. 2011. *The God Market: How Globalization Is Making India More Hindu*. Albany: NYU Press.

———. 2018. "Hindutva's Science Envy." *Frontline*, August 31, 2016.

Nandy, Ashis. 1988. *Science, Hegemony, and Violence: A Requiem for Modernity*. Delhi: Oxford University Press.

————. 1995. *Alternative Sciences: Creativity and Authenticity of Two Indian Scientists*, 2nd edition. Oxford: Oxford University Press.

———— 2002. *Time Warps: Silent and Evasive Pasts in Indian Politics and Religion*. New Brunswick, NJ: Rutgers University Press.

Nandy, Ashis, Shikha Trivedy, and Achyut Yagnik. 1995. *Creating a Nationality: The Ramjanmabhumi Movement and Fear of the Self*. New York: Oxford University Press.

Narain, Priyanka P. 2008. "Informal Hindu Alliance Starts Fussing over Setusamudram." *Live Mint*, July 7, 2008.

Narang, Ankita, Rishi Das Roy, Amit Chaurasia, Arijit Mukhopadhyay, Mitali Mukerji, and Debasis Dash. 2010. "IGVBrowser: A Genomic Variation Resource from Diverse Indian Populations." *Database: The Journal of Biological Databases and Curation*, baq022. doi:10.1093/database/baq022.

Narasimhan, T. E. 2014. "Political Battle Rages on Setusamudram Project." *Business Standard*, April 7, 2014.

Narasimhan, Vagheesh M., et al. 2018. "The Genomic Formation of South and Central Asia." Cold Spring Harbor Laboratory, *bioR$_x$iv: The Preprint Server for Biology*. https://doi.org/10.1101/292581.

Naresh, Kumar, Dubey Mukesh, and Agarwal Vivek. 2013. "Rudraksha: A Review on Mythological, Spiritual and Medicinal Importance." *Global Journal of Research on Medicinal Plants and Indigenous Medicine* 2, no. 1: 65–72.

Narrain, Arvind. 2004. *Queer: Despised Sexuality, Law, and Social Change*. Bangalore: Books for Change.

Narrain, Arvind, and Gautam Bhan. 2005. *Because I Have a Voice: Queer Politics in India*. Delhi: Yoda.

Narula, Smiti. 2001. "Caste Discrimination." In *Exclusion: A Symposium on Caste, Race, and the Dalit Question*, December 2001. www.india-seminar.com/2001/508/508%20 smita%20narula.htm.

Nash, Catherine. 2004. "Genetic Kinship." *Cultural Studies* 18, no. 1 (January 2004): 1–33.

————. 2012. "Gendered Geographies of Genetic Variation: Sex, Power and Mobility in Human Population Genetics." *Gender, Place, and Culture* 19, no. 4: 409–28.

Natrajan, Balmurli, and Suraj Jacob. 2018. "Putting Indian Food Habits in Their Place: 'Provincializing' Vegetarianism." *Economic and Political Weekly* 53, no. 9 (March 3): 54–64.

Nayak, Preeti. 2014. "The Three Ms of Commercial Surrogacy in India: Mother, Money, and Medical Market." In *Global and Transnational Surrogacy in India: Outsourcing Life*, edited by Sayantini Dasgupta and Shamita Dasgupta. Lanham: Lexington Books: 1–22.

Nayar, Pramod. K. 2008. *Postcolonial Literature: An Introduction*. New Delhi: Pearson Longman.

Naz Foundation v. Government of NCT Delhi. Delhi High Court of India at New Delhi, WP© No. 7455/2001, July 2, 2009.

NDTV. 2011. "Vaastu Will Play Important Role for Bangalore Metro Launch: Report." *NDTV.com*, October 10, 2011. www.ndtv.com/bangalore-news/vaastu -will-play-important-role-for-bangalore-metro-launch-report-565384.

———. 2014. "Narendra Modi Targets PM at Rally in Imphal." *NDTV.com*, February 8, 2014. www.ndtv.com/elections/article/india/narendra-modi-targets-pm-at-rally -in-imphal-says-nido-s-death-a-national-shame-480806.

Neelakantan, Shailaja, and Radha Udayakumar Ganesh Kumar. 2016. "Swaraj Takes to Twitter to Defend New Surrogacy Laws." *Times of India*, September 14, 2016.

Nehru, Jawaharlal. 1947. "Tryst with Destiny." *The Guardian*, April 30, 2007. www .theguardian.com/theguardian/2007/may/01/greatspeeches.

———. 1959. "India Today and Tomorrow." Azad Memorial Lectures, New Delhi, February 22, 1959. In *Jawaharlal Nehru Selected Speeches*, volume 4. New Delhi: Indian Council for Cultural Relations.

———. 1989. *The Discovery of India*, new edition. New Delhi: Oxford University Press.

Nelkin, Dorothy, and Susan Lindee. 2004. *The DNA Mystique: The Gene as a Cultural Icon*. Ann Arbor: University of Michigan Press.

Nelson, Alondra. 2016. *The Social Life of DNA: Race, Reparations, and Reconciliation after the Genome*. Boston: Beacon Press.

Nelson, Dean. 2009. "Hindu Guru Claims Homosexuality Can Be 'Cured' by Yoga." *Daily Telegraph*, July 8, 2009. www.telegraph.co.uk/news/worldnews/asia/india /5780028/Hinduguru-claims-homosexuality-can-be-cured-by-yoga.html.

New York Times. 2014. Editorial. "India's Public Health Crisis." *New York Times*, October 17, 2014.

———. 2017. Editorial. "Choking on Air in New Delhi." *New York Times*, November 12, 2017.

Nietzsche, Friedrich. 2000. *Basic Writings of Nietzsche*. Translated by Walter Kaufmann. New York: The Modern Library.

Niro, Brian. 2003. *Race*. Transitions. New York: Palgrave Macmillan.

Noble, David. 1992. *A World without Women: The Christian Clerical Culture of Western Science*. New York: Knopf.

Nongbri, B. 2013. *Before Religion*. New Haven, CT: Yale University Press.

Noorani, A. G. 1989. "The Babri Masjid–Ram Janmabhoomi Question." *Economic and Political Weekly* 24, no. 44/45: 2461–66.

O'Connor, Ashling. 2007. "Lord Ram Row Dredges Up Religious Fury." *The Times* (London), September 15, 2007.

Oikkonen, Venla. 2014. "Kennewick Man and the Evolutionary Origins of the Nation." *Journal of American Studies* 48, no. 1: 275–90.

Omvedt, Gail. 1995. *Dalit Visions: The Anti-Caste Movement and the Construction of an Indian Identity*. New Delhi: Orient Longman.

———. 2001a. "The UN, Racism and Caste I." *The Hindu*, April 9, 2001.

———. 2001b. "The UN, Racism and Caste II." *The Hindu*, April 10, 2001.

———. 2008. *Seeking Begumpura: The Social Vision of Anticaste Intellectuals*. New Delhi: Navayana.

Ong, Aihwa, and Nancy Chen. 2010. *Asian Biotech: Ethics and Communities of Fate*. Durham, NC: Duke University Press.

Osuri, Goldie. 2011. "Transnational Bio/Necropolitics: Hindutva and Its Avatars (Australia/India)." *Somatechnics* 1, no. 1: 138–60.

Oza, Rupal. 2007. "The Geography of Hindu Right-Wing Violence in India." In *Violent Geographies: Fear, Terror, and Political Violence*, edited by Derek Gregory and Allan Pred. New York: Routledge: 159–80.

Ozeki, Ruth. 2013. *A Tale for the Time Being*. New York: Viking Press.

Paddayya, K. 2012. "The Ramayana Controversy Again." *Bulletin of the Deccan College Research Institute* 72/73: 347–49.

Padhy, S. 2013. "Pancha Yajnya (Five Sacrifices): The Scientific Philosophy of Human Ecological Responsibility since the Vedic Age: A Review." *Journal of Biodiversity* 4, no. 1: 25–44.

Pálsson, G., 2007. "How Deep Is the Skin? The Geneticization of Race and Medicine." *BioSocieties* 2, no. 2: 257–72.

Pande, Amrita. 2009a. "Not an 'Angel,' Not a 'Whore': Surrogates as 'Dirty' Workers in India." *Indian Journal of Gender Studies* 16: 141–73.

———. 2009b. "'It May Be Her Eggs, But It's My Blood': Surrogates and The Every-day Forms of Kinship in India." *Qualitative Sociology* 32: 379–97.

———. 2010. "Commercial Surrogacy in India: Manufacturing a Perfect Mother-Worker." *Signs* 35, no. 4 (Summer): 969–92.

———. 2014a. *Wombs in Labor: Transnational Commercial Surrogacy in India*. New York: Columbia University Press.

———. 2014b. "The Power of Narratives: Negotiating Commercial Surrogacy in India." In *Global and Transnational Surrogacy in India: Outsourcing Life*, edited by Sayantini Dasgupta, and Shamita Dasgupta. Lanham, MD: Lexington Books: 87–106.

Panikkar, K. N. 2001. "Outsider as Enemy." *Frontline* 18, no. 1 (January 6–19). www .the-hindu.com/fline.

Pannu, S. P. S. 2014. "Narendra Modi Government May Dismantle the Planning Commission." *Business Today*, June 20, 2014. http://businesstoday.intoday.in /story/narendra-modi-govt-may-dismantle-the-planning-commission/1/207422 .html.

Pant, N. 1997. "Facilitating Genocide: Women as Fascist Educators in the Hindutva Movement." *Ghadar* 1, no. 1 (May 1). www.proxsa.org/ resources/ghadar/ghadar .html.

Parajuli, Pramod. 2001. "Learning from Ecological Ethnicities: Toward a Plural Political Ecology of Knowledge." In *Indigenous Traditions and Ecology*, edited by J. A. Grim. Cambridge, MA: Harvard University Press.

Parliament of India. 2017. "Report No. 102 on the Surrogacy (Regulation) Bill 2016." www.prsindia.org/uploads/media/Surrogacy/SCR-%20Surrogacy%20Bill,%20 2016.pdf.

Patel, Geeta. 2017. *Risky Bodies and Techno-Intimacy: Reflections on Sexuality, Media, Science*. Seattle: University of Washington Press.

Patel, Ishwarbhai, ed. 1984. *Science and the Vedas*. Bombay: Somaiya Publications.

Patrao, Michael. 2013. "Religion and Rocket Science in India." *The Media Project*.: http://themediaproject.org/article/religion-and-rocket-science-india?page=full.

Pattanaik, Devdutt. 2002. *The Man Who Was a Woman and Other Queer Stories from Hindu Lore*. New York: Harrington Park Press.

PBS Home Video. 2002. *The Journey of Man: The Story of the Human Species*. Directed by Clive Maltby. Producer: Tigress Productions. Hosted by Dr. Spencer Wells. PBS Home Video.

Peer, Basharat. 2014. "Being Muslim under Narendra Modi." *New York Times*, April 18, 2014.

Perur, Srinath. 2013. "The Origins of Indians: What Our Genes Are Telling Us." *Fountain Ink*, December 13, 2013. http://fountainink.in/?p=4669.

Pfeil, Fred. 1994. "No Basta Teorizar: In-Difference to Solidarity in Contemporary Fiction, Theory, and Practice." In *Scattered Hegemonies: Postmodernity and Transnational Feminist Practices*, edited by Inderpal Grewal and Caren Kaplan. Minneapolis: University of Minnesota Press: 197–230.

Phalkey, Jahnavi. 2013. "Introduction: Science, History, and Modern India." *Isis* 104, no. 2: 330–36.

Philip, Kavita. 2004. *Civilizing Natures: Race, Resources, and Modernity in Colonial South India*. New Brunswick, NJ: Rutgers University Press.

Philip, Kavita, Lilly Irani, and P. Dourish. 2012. "Postcolonial Computing: A Tactical Survey." *Science, Technology, and Human Values* 37, no. 1: 3–29.

Pinto, Ambrose. 2001a. "Caste Is a Variety of Race." *The Hindu*, March 24, 2001.

———. 2001b. "UN Conference against Racism: Is Caste Race?" *Economic and Political Weekly* 36, no. 30 (July 28–August 3): 2817–20.

Pinto, Sarah. 2014. *Daughters of Parvati: Women and Madness in Contemporary India*. Philadelphia: University of Pennsylvania Press.

Pirta, Raghubir Singh. 2009. "Biological and Ecological Bases of Behaviour." In *Psychology in India*, volume 1, *Basic Psychological Processes and Human Development*, edited by Girishwar Misra. Delhi: Longman: 1–67.

Pollock, Anne. 2014. "Places of Pharmaceutical Knowledge-Making: Global Health, Postcolonial Science, and Hope in South African Drug Discovery." *Social Studies of Science* 44, no. 6: 848–73.

———. 2016. "Queering Endocrine Disruption." In *Object-Oriented Feminism*, edited by Katherine Behar. Minneapolis: University of Minnesota Press.

Pollock, Anne, and Banu Subramaniam. 2016. "Resisting Power, Retooling Justice: Promises of Feminist Postcolonial Technosciences." *Science, Technology, and Human Values* 41, no. 6: 951–66.

Prakash, Gyan. 1990. "Writing Post-Orientalist Histories of the Third World Perspectives from Indian Historiography." *Comparative Studies in Society and History* 32, no. 2: 383–408.

———. 1999. *Another Reason: Science and the Imagination of Modern India*. Princeton, NJ: Princeton University Press.

Prasad, Amit. 2008. "Science in Motion: What Postcolonial Science Studies Can Offer." *Electronic Journal of Communication Information and Innovation in Health* (RECIIS) 2, no. 2 (July–December): 35–47.

———. 2014. *Imperial Technoscience: Transnational Histories of MRI in the United States, Britain, and India*. Cambridge, MA: MIT Press.

Prasad, Rajendra. 1993. "Debiprasad Chattopadhyaya." *Social Scientist* 21, nos. 5/6 (May–June 1993): 102–5.

Prasher, Bhavana, Greg Gibson, and Mitali Mukerji. 2016. "Genomic Insights into Ayurvedic and Western Approaches to Personalized Medicine." *Journal of Genetics* 95, no.1: 209–28.

Prasher, Bhavana, Binuja Varma, Arvind Kumar, Bharat Krushna Khuntia, Rajesh Pandey, Ankita Narang, Pradeep Tiwari, Rintu Kutum, Debleena Guin, Ritushree Kukreti, Debasis Dash, TRISUTA Ayurgenomics Consortium, and Mitali Mukerji. 2017. "Ayurgenomics for Stratified Medicine: TRISUTRA Consortium Initiative across Ethnically and Geographically Diverse Populations." *Journal of Ethnopharmacology* 197 (February 2): 274–93.

Pratt, Mary Louise. 1992. *Imperial Eyes: Travel Writing and Transculturation*. London: Routledge.

Pulla, P. 2014. "Searching for Science in India's Traditional Medicine." *Science* 346, no. 6208: 410–10.

Punwani, Jyoti. 2014. "Myths and Prejudices about 'Love Jihad.'" *Economic and Political Weekly* 49, no. 42 (October 18).

Puri, Jyoti. 2002a. *Woman, Body, Desire in Post-colonial India: Narratives of Gender and Sexuality*. New York: Routledge.

———. 2002b. "Concerning Kamasutras: Challenging Narratives of History and Sexuality." *Signs: Journal of Women in Culture and Society* 27, no. 3: 603–39.

———. 2012. "Sexualizing the State: Sodomy, Civil Liberties, and the Indian Penal Code." In *Contesting Nation: Gendered Violence in South Asia: Notes on the Postcolonial Present*, edited by Angana Chattrji and Lubna Nazir Chaudhry. New Delhi: Zubaan Press.

———. 2016. *Sexual States: Governance and the Struggle over the Antisodomy Law in India*. Durham, NC: Duke University Press.

Purkayastha, Shramana Das. 2014. "Against the Order of Nature? Postcolonial State, Section 377 and the Homosexual Subject." "Special Issue on LGBT and Queer Studies," *Rupkatha Journal on Interdisciplinary Studies in Humanities* 6, no. 1: 120–30.

Quigley, Declan. 2000. *The Interpretation of Caste*. New York: Oxford India Paperbacks.

Rabinow, Paul, and Nikolas Rose. 2006. "Thoughts on the Concept of Biopower Today." *BioSocieties* 1, no. 2: 195–217.

Rabinowitz, Abby. 2016. "The Trouble with Renting a Womb." *Guardian*, April 28, 2016.

Rahman, Masseeh. 2014. "Indian Prime Minster Claims Genetic Science Existed in Ancient Times." *The Guardian*, October 28, 2014.

Raina, Dhruv. 1996. "Reconfiguring the Center: The Structure of Scientific Exchanges between Colonial India and Europe." *Minerva* 34, no. 2: 161–76.

———. 1999. "From West to Non-West? Basalla's Three-Stage Model Revisited." *Science as Culture* 8: 497–516.

———. 2003. *Images and Contexts: The Historiography of Science and Modernity in India.* Delhi: Oxford University Press.

Raina, Dhruv, and Irfan Habib. 2006. *Domesticating Modern Science: A Social History of Science and Culture in Colonial India.* New Delhi: Tulika.

Raj, Kapil. 2001. "Colonial Encounters and the Forging of New Knowledge and National Identities: Great Britain and India, 1760–1850." *Osiris* 15: 119–34.

———. 2006. *Relocating Modern Science: Circulation and the Construction of Scientific Knowledge in South Asia and Europe.* Delhi: Permanent Black.

———. 2013. "Beyond Postcolonialism . . . and Postpositivism: Circulation and the Global History of Science." *Isis* 104, no. 2: 337–47.

Rajagopal, Arvind. 2000. "Hindu Nationalism in the US: Changing Configurations of Political Practice." *Ethnic and Racial Studies* 23, no. 3: 467–96.

Rajagopal, Krishnadas. 2016. "Five-Judge Constitution Bench to Take a Call on Section 377." *The Hindu*, February 2, 2016.

Rajshekar, V. T. 2009. *Dalit: The Black Untouchables of India.* Atlanta, GA: Clarity Press.

Ramachandra, Komala. 2014. "The International Interest." *Caravan*, August 2014. www.caravanmagazine.in/perspectives/international-interest.

Ramachandran, R. 2001. "New Genetic Evidence for the Origin of Castes Indicates That the Upper Castes Are More European than Asian." *Frontline* 18, no. 12 (June 9).

Ramadurai, Chitra. 2015. "India's Temples of Sex." *BBC Travel*, October 7, 2015. www.bbc.com/travel/story/20150921-indias-temples-of-sex.

Raman, J. S. 2000. "The Bomb Club." *Hindustan Times*, November 11, 2000.

Ramanujan, A. K. 1989. "Is There an Indian Way of Thinking? An Informal Essay." *Contributions to Indian Sociology* 23, no. 1: 41–58.

———. 1991. "Three Hundred Rāmāyaṇas: Five Examples and Three Thoughts on Translation." In *Many Rāmāyaṇas: The Diversity of a Narrative Tradition in South Asia,* edited by Paula Richman. Berkeley: University of California Press: 22–48.

Ramaswamy, Sumathi. 2010. *The Goddess and the Nation: Mapping Mother India.* Durham, NC: Duke University Press.

Ramberg, Lucinda. 2014. *Given to the Goddess: South Indian Devadasis and the Sexuality of Religion.* Durham, NC: Duke University Press.

Ramesh, R. 2004. "Setusamudram Shipping Canal Project and the Unconsidered Risk Factors: Can It Withstand Them?" www.elaw.org/system/files/Setusamudram_Shipping_Canal_Project_Final.pdf.

———. 2005. "Will to Disaster: Post-Tsunami Technical Feasibility of Sethusamudram Project." *Economic and Political Weekly* 41, no. 26 (June 25–July 1): 2648–51+2653.

Rangarajan, Swarnalatha. 2009. "Ecological Dimensions of the Ramayana: A Conversation with Paula Richman." *The Trumpeter* 25, no. 1: 22–33.

Rao, Anupama. 2009. *The Caste Question: Dalits and the Politics of Modern India*. Berkeley: University of California Press.

Rao, D. V. Subba, K. Srinivasa Rao, C. S. P. Iyer, and P. Chittibabu. 2008. "Possible Ecological Consequences from the Setu Samudram Canal Project, India." *Marine Pollution Bulletin* 56, no. 2 (February): 170–86.

Rawat, B. 2000. "Advani Tribute in Concrete to Iron Man." *Daily Telegraph*, October 31, 2000.

Ray, Udayan. 1998. "Houses of the Holy." *Money Magazine*, August 12, 1998:. www.outlookmoney.com/printarticle.aspx?85995.

Raza, Danish. 2014. "Saffronising Texbooks: Where Myth and Dogma Replace History." *Hindustan Times*, December 8, 2014.

Reardon, Jenny. 2009. *Race to the Finish: Identity and Governance in an Age of Genomics*. Princeton, NJ: Princeton University Press.

———. 2017. *The Postgenomic Condition: Ethics, Justice, and Knowledge after the Genome*. Chicago: University of Chicago Press.

Reardon, Jenny, and Kim TallBear. 2012. "'Your DNA Is Our History': Genomics, Anthropology, and the Construction of Whiteness as Property." *Current Anthropology* 53, no. S5: S233-S245.

Reardon, Jenny, Jacob Metcalf, Martha Kenney, and Karen Barad. 2015. "Science and Justice: The Trouble and the Promise." *Catalyst: Feminism, Theory, Technoscience* 1, no. 1: 1–48.

Reddy, B. Mohan, B. T. Langstieh, Vikrant Kumar, T. Nagaraja, A. N. S. Reddy, Aruna Meka, A. G. Reddy, K. Thangaraj, and Lalji Singh. 2007. "Austro-Asiatic Tribes of Northeast India Provide Hitherto Missing Genetic Link between South and Southeast Asia." *PLoS One* 2, no. 11: e1141.

Reddy, Deepa S. 2005. "The Ethnicity of Caste." *Anthropological Quarterly* 78, no. 3: 543–84.

Rediff India. 2007. "Lord Ram Is an Imaginary Character: Karunanidhi." *Rediff.com* Setpember 15, 2007. www.rediff.com//news/2007/sep/15setu.htm.

Reich, David, Kumaraamy Thangaraj, Nick Patterson, Alkes L. Price, and Lalji Singh. 2009. "Reconstructing Indian Population History." *Nature* 461, no. 7263: 489–94.

Richerson, Peter, and J. Robert Boyd. 2005. *Not by Genes Alone: How Culture Transformed Human Evolution*. Chicago: University of Chicago Press.

Ridley, Louse. 2014. "Indian Prime Minister Narendra Modi Says 'May the Force Be with You' with Hugh Jackman in Surreal New York Visit." *Huffington Post*, September 29, 2014. www.huffingtonpost.co.uk/2014/09/29/narendra-modi-new-york-madison-square-garden-john-oliver_n_5898584.html.

Roberts, Dorothy. 2009. "Race, Gender, and Genetic Technologies: A New Reproductive Dystopia?" *Signs* 34, no. 4: 783–803.

Roberts, Elizabeth. 2012. *God's Laboratory: Assisted Reproduction in the Andes*. Berkeley: University of California Press.

———. 2016. "Gods, Germs, and Petri Dishes: Toward a Nonsecular Medical Anthropology." *Medical Anthropology* 35, no. 3: 209–19. doi 10.1080/01459740.2015.1118100.

Rocha, Leon Antonio. 2011. "Scientia Sexualis versus Ars Erotica: Foucault, van Gulik, Needham." *Studies in History and Philosophy of Biological and Biomedical Sciences* 42, no. 3: 328–43.

Rodrigues, Sudarshan. 2007. "Review of the Environmental Impacts of the Setusamudram Ship Canal Project." *Indian Ocean Turtle Newsletter* no. 6 (July).

Rogers, Alan. 2001."Order Emerging from Chaos in Human Evolutionary Genetics." *Proceedings of the National Academy of Sciences* 98, no. 3: 779–80.

Roosth, Sophia. 2017. *Synthetic: How Life Got Made*. Chicago: University of Chicago Press.

Rose, Nikolas. 2006. *The Politics of Life Itself: Biomedicine, Power, and Subjectivity in the Twenty-First Century*. Princeton, NJ: Princeton University Press.

Rose, Nikolas, and Carlos Novas. 2004. "Biological Citizenship." In *Global Assemblages: Technology, Politics, and Ethics as Anthropological Problems*, edited by Aihwa Ong and Stephen J. Collier. Oxford: Blackwell: 439–63.

Roughgarden, Joan. 2004. *Evolution's Rainbow: Diversity, Gender, and Sexuality in Nature and People*. Los Angeles: University of California Press.

———. 2005. "The Myth of Sexual Selection." *California Wild: The Magazine of the California Academy of Sciences*, Summer.

———. 2009. *The Genial Gene: Deconstructing Darwinian Selfishness*. Berkeley: University of California Press.

Roy, Arundhati. 1997. *The God of Small Things*. New York: Random House.

———. 2000. "The End of Imagination." In *New Nukes: India, Pakistan, and Global Nuclear Disarmament*, edited by Praful Bidwai and Achin Vanaik. Oxford: Signal Books: xix–xxix.

———. 2016. *The End of Imagination*. Chicago: Haymarket Books.

Roy, Deboleena. 2017. "Reproductive Justice and Transplacental Politics." Paper presented at the National Women's Studies Association Annual Meeting, Baltimore, MD, November 17, 2017.

Roy, Srirupa. 2007. *Beyond Belief: India and the Politics of Postcolonial Nationalism*. Durham, NC: Duke University Press.

Rudrappa, Sharmila. 2010. "Making India the 'Mother Destination': Outsourcing Labor to Indian Surrogates." In *Gender and Sexuality in the Workplace*, edited by Christine L. Williams and Kirsten Dellinger. Bingley: Emerald Group Publishing: 253–85.

———. 2015. *Discounted Life: The Price of Global Surrogacy in India*. New York: New York University Press.

Rushdie, Salman. 1981. *Midnight's Children*. New York: Knopf.

———. 1991. "Excerpts from Rushdie's Address: 1000 Days 'Trapped Inside a Metaphor.'" *New York Times*, December 12, 1991.

———. 2015. *Two Years Eight Months and Twenty-Eight Nights*. New York: Random House

Sabir, Sharjeel. 2003. "Chimerical Categories: Caste, Race, and Genetics." *Developing World Bioethics* 3, no. 2: 170–77.

Sacco, Katherine, and Hilary Agro. 2017. "Entanglement's Limits: An Interview with Elizabeth F. S. Roberts." Dialogues, *Cultural Anthropology* website, March 12, 2–18.

Sachitanand, Rahul. 2017. "Why Companies Like HUL, Patanjali, Dabur Are Taking a Crack at the Market for Ayurvedic and Herbal Products." *Economic Times*, October 15, 2017.

Sagar. 2017. "Down the Drain: How the Swachh Bharat Mission Is Heading for Failure." *Caravan: A Journal of Politics and Culture*, May 1, 2017.

Sah, N. K., S. N. P. Singh, S. Sahdev, S. Banerji, V. Jha, Z. Khan, and S. E. Hasnain. 2005. "Indian Herb 'Sanjeevani'(*Selaginella bryopteris*) Can Promote Growth and Protect against Heat Shock and Apoptotic Activities of Ultra Violet and Oxidative Stress." *Journal of Biosciences* 30, no. 4: 499–505.

Sahoo, S., A. Singh, G. Himabindu, J. Banerjee, T. Sitalaximi, S. Gaikwad, R. Trivedi, P. Endicott, T. Kivisild, M. Metspalu, and R. Villems. 2006. "A Prehistory of Indian Y Chromosomes: Evaluating Demic Diffusion Scenarios." *Proceedings of the National Academy of Sciences of the United States of America* 103, no. 4: 843–48.

Said, Edward. 1993. *Culture and Imperialism*. New York: Vintage Books (Random House).

Salam, Ziya Us. 2017. "Scars of Memory." *Frontline*, July 21, 2017.

Sama. 2012a. *Birthing a Market: A Study on Commercial Surrogacy*. New Delhi: Sama—Resource Group for Women and Health.

———. 2012b. *Regulation of Surrogacy in Indian Context*. New Delhi: Sama—Resource Group for Women and Health.

Sampath, G. 2017. "Let a Billion 'Uttam Santatis' Bloom." *The Hindu*, May 13, 2017.

Sanders, Douglas. 2009. "377 and the Unnatural Afterlife of British Colonialism in Asia." *Asian Journal of Comparative Law* 4, no. 1: 1–49.

Sangari, Kumkum, and Sudesh Vaid, eds. 1989. *Recasting Women: Essays in Colonial History*. New Delhi: Kali for Women in India.

Saravanan, S. 2010. "Transnational Surrogacy and Objectification of Gestational Mothers." *Economic and Political Weekly* 45, no. 16 (April 17–23): 26–29.

Sarkar, Soumya. 2018. "India Has More Environmental Conflicts Than Any Other Country in the World." *The Hindu*, June 2, 2018.

Sarkar, Tanika. 1994. "Women, Community, and Nation: A Historical Trajectory for Hindu Identity Politics." In *Appropriating Gender: Women's Activism and Politicized Religion in South Asia*, edited by Patricia Jeffery and Amrita Basu. London: Routledge: 89–104.

———. 2001. *Hindu Wife, Hindu Nation: Community, Religion, and Cultural Nationalism*. Bloomington: Indiana University Press.

———. 2008. *Visible Histories, Disappearing Women: Producing Muslim Womanhood in Late Colonial Bengal*. Durham, NC: Duke University Press.

Savary, Luzia. 2014. "Vernacular Eugenics? *Santati-śāstra*' in Popuar Hindi Advisory Literature (1900–1940)." *South Asia: Journal of South Asian Studies* 37, no. 3: 381–97.

Saxena, R., D. Saleheen, L. F. Been, M. L. Garavito, T. Braun, A. Bjonnes, R. Young, W. K. Ho, A. Rasheed, P. Frossard, and X. Sim. 2013. "Genome-Wide Association Study Identifies a Novel Locus Contributing to Type 2 Diabetes Susceptibility in Sikhs of Punjabi Origin from India." *Diabetes* 62, no. 5: 1746–55.

Schiebinger, Londa. 1989. *The Mind Has No Sex? Women in the Origins of Modern Science.* Cambridge, MA: Harvard University Press.

———. 1993. *Nature's Body: Gender in the Making of Modern Science.* New Brunswick, NJ: Rutgers University Press.

———. 2004. *Plants and Empire: Colonial Bioprospecting in the Atlantic World.* Cambridge, MA: Harvard University Press.

———. 2005. "Forum Introduction: The European Colonial Science Complex." *Isis* 96, no. 1: 52–55.

Schiebinger, Londa, and Claudia Swan, eds. 2005. *Colonial Botany: Science, Commerce, and Politics in the Early Modern World.* Philadelphia: University of Pennsylvania Press.

Schultz, Kai, and Suhasini Raj. 2017. "'We Are Afraid of Christmas': Tensions Dampen Holiday in India." *New York Times*, December 24, 2017.

Scott, Walter. 2014. "Gadkari to Conduct Aerial Inspection of Setusamudram Project Area." *The Hindu*, November 3, 2014.

Sehgal, M. 2015. "Defending the Nation: Militarism, Women's Empowerment, and the Hindu Right." In *Border Politics: Social Movements, Collective Identities, and Globalization*, edited by Nancy Naples and Jennifer Mendez. New York: NYU Press: 60–94.

Sengupta, Anasuya. 2016. "Erasing History: What the Battle over California's Textbooks Really Means." *The Wire*, May 4, 2016.

Seth, Suman. 2009. "Putting Knowledge in Its Place: Science, Colonialism, and the Postcolonial." *Postcolonial Studies* 12, no. 4: 373–88.

Sethi, Manisha. 2002. "The Avenging Angel and the Nurturing Mother: Women and Hindu Nationalism." *Economic and Political Weekly* 37, no. 16: 1545–52.

Shah, Svati P. 2015. "Queering Critiques of Neoliberalism in India: Urbanism and Inequality in the Era of Transnational 'LGBTQ' Rights." *Antipode* 47, no. 3: 635–51.

Shandilya, Krupa. 2016. "(In)visibilities: Homosexuality and Muslim Identity in India after Section 377." *Signs* 42, no. 2: 459–84.

Sharma, Mukul. 2009. "Passages from Nature to Nationalism: Sunderlal Bahuguna and Tehri Dam Opposition in Garhwal." *Economic and Political Weekly* 44, no. 8 (February 21): 35–42.

———. 2011. *Green and Saffron: Hindu Nationalism and Indian Environmental Politics.* Ranikhet: Orient Blackswan.

Sharma, Sanchita. 2017. "No More Rent-a-Womb: Why India Needs to Regulate Surrogacy." *Hindustan Times*, August 24, 2017.

Shea, Elizabeth Parthenia. 2008. *How the Gene Got Its Groove: Figurative Language, Science and the Rhetoric of the Real.* Albany NY: SUNY Press.

Shiva, Vandana. 1988. "Reductionist Science as Epistemological Violence." In *Science, Hegemony, and Violence: A Requiem for Modernity*, edited by Ashis Nandy. Delhi: Oxford University Press.

———. 1989. *Staying Alive: Women, Ecology, and Development*, London: Zed Books.

———. 1993. *Monocultures of the Mind: Perspectives on Biodiversity and Biotechnology*. New York: Palgrave Macmillan.

———. 2016. *Biopiracy: The Plunder of Nature and Knowledge*. Berkeley, CA: North Atlantic Books.

Shivashankar, B. V. 2015. "Vaastu Expert Made Telangana Government Adviser." *Times of India*, February 23, 2015.

Shotwell, Alexis. 2016. *Against Purity: Living Ethically in Compromised Times*. Minneapolis: University of Minnesota Press.

Shrivastava, Saurabh RamBihariLal, Prateek Saurabh Shrivastava, and Jegadeesh Ramasamy. 2015. "Mainstreaming of Ayurveda, Yoga, Naturopathy, Unani, Siddha, and Homeopathy with the Health Care Delivery System in India." *Journal of Traditional Complementary Medicine* 5, no. 2: 116–18.

Silva-Zolezzi, I., A. Hidalgo-Miranda, J. Estrada-Gil, J. C. Fernandez-Lopez, L. Uribe-Figueroa, A. Contreras, E. Balam-Ortiz, L. del Bosque-Plata, D. Velazquez-Fernandez, C. Lara, and R. Goya. 2009. "Analysis of Genomic Diversity in Mexican Mestizo Populations to Develop Genomic Medicine in Mexico." *Proceedings of the National Academy of Sciences* 106, no. 21: 8611–16.

Singh, Madhur. 2007. "India's Debate Over Sacred Geography." *Time Magazine*, September 18, 2007.

Singh, Manmohan. 2005 "Prime Minster Address at the Commencement of Work on the Setusamudram Ship Channel Project." https://web.archive.org/web /20071118013715/http://pmindia.nic.in/speech/content.asp?id=511.

Singh, R. D., N. Haridas, F. D. Shah, J. B. Patel, S. N. Shukla, and P. S. Patel. 2014. "Gene Polymorphisms, Tobacco Exposure and Oral Cancer Susceptibility: A Study from Gujarat, West India." *Oral Diseases* 20, no. 1: 84–93.

Singh, Rama S. 2000. "The Indian Caste System." In *Thinking about Evolution: Historical, Philosophical, and Political Perspectives*, edited by Rama S. Singh, Costas B. Krimbas, Diane B. Paul, and John Beatty. Cambridge, UK: Cambridge University Press: 152–83.

Singh, T. 1998. "Beware the Nuclear Yogis." *India Today*, June 1, 1998.

Singh, Vandana. 2008. "A Speculative Manifesto." In *The Woman Who Thought She Was a Planet and Other Stories*. New Delhi: Zubaan: 200–203.

Sinha, Subir, S. Gururani, and B. Greenberg. 1997. "The 'New Traditionalist' Discourse of Indian Environmentalism." *Journal of Peasant Studies* 24, no. 3: 65–99.

Sivaramakrishnan, K. 2011. "Thin Nationalism: Nature and Public Intellectualism in India." *Contributions to Indian Sociology* 45, no. 1: 85–111.

Skinner, David. 2006. "Racialized Futures: Biologism and the Changing Politics of Identity." *Social Studies of Science* 36, no. 3: 459–88.

Spargo, Tasmin. 1999. *Foucault and Queer Theory*. Postmodern Encounters. Cambridge, UK: Icon Books.

Spivak, Gayatri. 1999. *A Critique of Postcolonial Reason*. Cambridge, MA: Harvard University Press.

Sreenivas, Deepa. 2010. *Sculpting the Middle Class: History, Masculinity, and the Amar Chitra Katha*. New York: Routledge.

Srinivas, M. N. 2000. *Caste: Its Twentieth Century Avatar*. New Delhi: Penguin India.

Stanley, Oswin D. 2004. "Ecological Balance of Setusamudram Canal, India: Special Reference to Mangrove Ecosystem." *Journal of Coastal Development* 8, no. 1: 1–10.

Stepan, Nancy. 1982. *The Idea of Race in Science: Great Britain, 1800–1960*. Hamden, CT: Archon Books.

———. 1991. *The Hour of Eugenics: Race, Gender, and Nation in Latin America*. Ithaca, NY: Cornell University Press

Stephens, Philip. 2014. "India's Narendra Modi Joins the Great Power Game." *Financial Times*, November 20, 2014.

STHV (Science, Technology and Human Values). "Resisting Power, Retooling Justice: Feminist Postcolonial Technosciences," *Science, Technology, and Human Values* 41, no. 6 (special issue, November 2016).

Stoler, Ann Laura. 2002. *Carnal Knowledge and Imperial Power: Race and the Intimate in Colonial Rule*. Berkeley: University of California Press.

Subramaniam, Banu. 2000. "Archaic Modernities: Science, Secularism and Religion in Modern India." *Social Text* 18, no. 3: 67–86.

———. 2013. "Re-Owning the Past: DNA and the Politics of Belonging." In *Negotiating Culture: Heritage, Ownership, and Intellectual Property*, edited by Laetitia La Follette. Amherst: University of Massachusetts Press.

———. 2014. *Ghost Stories for Darwin: The Science of Variation and the Politics of Diversity*. Urbana: University of Illinois Press.

———. 2015a. "Colonial Legacies, Postcolonial Biologies: Gender and the Promises of Biotechnology." *Asian Biotechnology and Development Review* 17, no. 1 (March): 15–36.

———. 2015b. "Science and Postcolonialism." In *A Companion to the History of American Science*, edited by Montgomery and Mark Larger. Chichester: Blackwell Press.

———. 2017. "Recolonizing India: Troubling the Anti-colonial, Decolonial, Postcolonial. "Science Out of Feminist Theory, part 1: Feminism's Sciences," special issue of *Catalyst: Feminism, Theory, Technoscience* 3, no. 1: 10–13.

Sugirtharajah, R. S. 2018. *Jesus in Asia*. Cambridge, MA: Harvard University Press.

Sujatha, V. 2007. "Pluralism in Indian Medicine: Medical Lore as a Genre of Medical Knowledge." *Contributions to Indian Sociology* 41, no. 2: 169–202.

———. 2011. "Innovation within and between Traditions: Dilemma of Traditional Medicine in Contemporary India." *Science, Technology, and Society* 16, no. 2: 191–213.

Sujatha, V., and Leena Abraham, eds. 2012. *Medical Pluralism in Contemporary India*. New Delhi: Orient Blackswan, 2012.

Sunder Rajan, Kaushik. 2006. *Biocapital: The Constitution of Postgenomic Life*. Durham, NC: Duke University Press.

———. 2017. *Pharmocracy: Value, Politics, and Knowledge in Global Biomedicine*. Durham, NC: Duke University Press.

Sur, Abha. 2011. *Dispersed Radiance: Caste, Gender, and Modern Science in India*. New Delhi: Navanyana.

Sur, Abha, and Samir Sur. 2008. "In Contradiction Lies the Hope: Human Genome and Identity Politics." In *Tactical Biopolitics: Art, Activism, and Technoscience*, edited by Beatriz da Costa and Kavita Philip. Cambridge MA: MIT Press: 269–87.

Surendran, S. B. S. N.d. "Vaastu and Feng Shui." *Feng Shui Today for the Modern Professional*, https://www.fengshuitoday.com/vaastu-and-feng-shui/ (accessed September 20, 2018).

Surrana, Pingali. 2002. *The Sound of the Kiss, or The Story that Must Never be Told*. Translated by Velcheru Narayana Rao and David Shulman. New York: Columbia University Press.

Surrogacy (Regulation) Bill, 2016. www.prsindia.org/uploads/media/Surrogacy /Surrogacy%20(Regulation)%20Bill,%202016.pdf.

Swanson, K. W. 2014. *Banking on the Body*. Cambridge, MA: Harvard University Press.

Sylvester, Christine. 2006. "Bare Life as a Development/Postcolonial Problematic." *Geographical Journal* 172, no. 1: 66–77.

TallBear, Kim. 2014. "Standing with and Speaking as Faith: A Feminist-Indigenous Approach to Inquiry." *Journal of Research Practice* 10, no. 2: Article N17.

Tamang, Rakesh, Lalji Singh, and Kumarasamy Thangaraj. 2012. "Complex Genetic Origin of Indian Populations and Its Implications." *Journal of Biosciences* 37, no. 5: 911–19.

Tambe, Ashwini. 2009. *Codes of Misconduct: Regulating Prostitution in Late Colonial Bombay*. Minneapolis: University of Minnesota Press.

Tan, S. T., W. Scott, V. Panoulas, J. Sehmi, W. Zhang, J. Scott, P. Elliott, J. Chambers, and J. S. Kooner. 2014. "Coronary Heart Disease in Indian Asians." *Global Cardiology Science and Practice*, no. 1: 13–23.

Taneja, N. 2000. "Communalisation of Education in India: An Update." *Akhbàr: A Window on South Asia* 1. http://wzw.xitami.net/indowindow/akhbar.

Teo, Y. Y., X. Sim, R. T. Ong, A. K. Tan, J. Chen, E. Tantoso, K. S. Small, C. S. Ku, E. J. Lee, M. Seielstad, and K. S. Chia. 2009. "Singapore Genome Variation Project: A Haplotype Map of Three Southeast Asian Populations." *Genome Research* 19, no. 11: 2154–62.

Teresi, Dick. 2001. *Lost Discoveries: Ancient Roots of Modern Science—From the Babylonians to the Mayans*. New York: Simon and Schuster.

Terry, J. 2000. "'Unnatural Acts' in Nature: The Scientific Fascination with Queer Animals." *GLQ: A Journal of Lesbian and Gay Studies* 6, no. 2: 151–93.

Thangaraj, K., G. V. Ramana, and L. Singh. 1999. "Y-Chromosome and Mitochondrial DNA Polymorphisms in Indian Populations." *Electrophoresis* 20, no. 8: 1743–47.

Thapar, Romila. 1995. "The First Millennium B.C. in India," In *Recent Perspectives of Early Indian History*, edited by Romila Thapar. Bombay: Popular Prakashan: 87–150.

Tharoor, Shashi. 2004. "Globalization and the Human Imagination." *World Policy Journal* 21, no. 2 (Summer): 85–91.

Thatai, Mukund. 2018. "A Billion Candles: Is There an Indian Way of Doing Science?" *The Wire*, March 20. https://thewire.in/the-sciences/a-billion-candles-is-there-an -indian-way-of-doing-science.

Thomas, Renny. 2016. "Science, Religion, and Atheism in Contemporary India." In *Science and Religion: East and West*, edited by Yiftach Fehige. New York: Routledge: 140–57.

———. 2017. "Atheism and Unbelief among Indian Scientists: Towards an Anthropology of Atheism(s)." *Society and Culture in South Asia* 3, no. 1: 45–67.

Times of India. 2002. "Hanuman Bridge Is Myth: Experts." *Times of India*, October 19, 2002. http://timesofindia.indiatimes.com/india/Hanuman-bridge-is-myth-Experts/articleshow/25601383.cms.

———. 2007. "RSS, Bajrang Dal Workers Take to Streets." *Times of India*, September 13, 2007.

———. 2009. "BJP to Declare Ramsetu 'Heritage Monument' If It Comes to Power." *Times of India*, March 22, 2009.

———. 2017. "PM Modi Urges India Inc to Invest More in Ayurveda." *Times of India*, October 18, 2017. https://timesofindia.indiatimes.com/india/pm-modi-urges-india -inc-to-invest-more-in-ayurveda/articleshow/61125610.cms.

———. 2018. "Medical Bridge Course Widens Gulf." *Times of India*, February 6, 2018.

Tomalin, Emma. 2009. *Biodivinity and Biodiversity: The Limits to Religious Environmentalism for India*. Aldershot: Ashgate Publishing.

Traweek, Sharon. 1988. *Beamtimes and Lifetimes: The World of High Energy Physicists*. Cambridge, MA: Harvard University Press.

Tsing, Anna. 2015. *The Mushroom at the End of the World: On the Possibility of Life in Capitalist Ruins*. Princeton, NJ: Princeton University Press.

Underhill, P. A., N. M. Myres, S. Rootsi, M. Metspalu, L. A. Zhivotovsky, R. J. King, A. A. Lin, C. E. T. Chow, O. Semino, V. Battaglia, and I. Kutuev. 2010. "Separating the Post-Glacial Coancestry of European and Asian Y Chromosomes within Haplogroup R1a." *European Journal of Human Genetics* 18, no. 4: 479–84.

Vaastu International. N.d. Website. www.vaastuinternational.com/subject.html (accessed September 20, 2018).

Vaastuyogam. 2012. "Architect's Voice: Navin Ghorecha." *Vaastuyogam*, October 2012. www.vaastuyogam.com/wp-content/uploads/2012/10/architect_voice_mr_navin _ghorecha.pdf.

Vanaik, Achin. 1997. *The Furies of Indian Communalism: Religion, Modernity, and Secularization*. New York: Verso.

van der Veer, Peter. 1994. *Religious Nationalism: Hindus and Muslims in India*. Berkeley: University of California Press.

van Dijk, Michiel, and Virginie Mamadouh. 2011. "When Megaengineering Disturbs Ram: The Sethusamudram Ship Canal Project." In *Engineering Earth*, edited by Standley Brunn. Dordrecht: Springer: 297–310.

Vanita, Ruth. 2007. "'Living the Way We Want': Same-Sex Marriage in India." In *The Phobic and the Erotic: The Politics of Sexualities in Contemporary India*, edited by Brinda Bose and Subhabrata Bhattacharya. Calcutta: Seagull Books: 342–65.

———, ed. 2013. *Queering India: Same-Sex Love and Eroticism in Indian Culture and Society*. New York: Routledge.

Vanita, Ruth, and Saleem Kidwai, eds. 2000. *Same-Sex Love in India: Readings in Indian Literature*. New York: St. Martin's Press.

Varshney, A. 2000. "Is India Becoming More Democratic?" *Journal of Asian Studies* 59, no. 1: 3–25.

Vemsani, Lavanya. 2014. "Genetic Evidence of Early Human Migrations in the Indian Ocean Region Disproves Aryan Migration/Invasion Theories." In *Sindhu Saraswathi Civilization: A Reappraisal*," edited by Nalini Rao. Louisiana: DK Print World and Nalanda International. 594–621.

Vendell, Dominic. 2014. "Jotirao Phule's Satyashodh and the Problem of Subaltern Consciousness." *Comparative Studies of South Asia, Africa and the Middle East* 34, no. 1: 52–66.

Venkatachalam, J., S. B. Abrahm, Z. Singh, P. Stalin, and G. R. Sathya. 2015. "Determinants of Patient's Adherence to Hypertension Medications in a Rural Population of Kancheepuram District in Tamil Nadu, South India." *Indian Journal of Community Medicine: Official Publication of Indian Association of Preventive and Social Medicine* 40, no. 1: 33–37.

Verran, Helen. 2001. *Science and an African Logic*. Chicago: University of Chicago Press.

———. 2002. "A Postcolonial Moment in Science Studies: Alternative Firing Regimes of Environmental Scientists and Aboriginal Landowners." *Social Studies of Science* 32, nos. 5–6: 729–62.

Vishnoi, Anubhuti, and Seema Chishti. 2014. "BJP's Muslim Score: 7 of 482 Fielded, No Winners." *Indian Express*, May 19, 2014.

Viswanath, Radha. 2017. *Ravana Leela: The One Who Forced God to Become Human*. New Delhi: Rupa Publications India.

Viswanathan, Shiv. 1997. *A Carnival for Science*. New Delhi: Oxford University Press.

———. 2001. "The Race for Caste: Prolegomena to the Durban Conference." *Economic and Political Weekly* 36, no. 27 (July 7–13): 2512–16.

Viswesaran, Kamala. 1994. *Fictions of Feminist Ethnography*. Minneapolis: University of Minnesota Press.

———. 2010. *Un/Common Cultures: Racism and the Rearticulation of Cultural Difference*, Durham, NC: Duke University Press.

Viswesaran, Kamala, Michael Witzel, Nandini Manjrekar, Dipta Bhog, and Uma Chakravarti. 2009. "The Hindutva View of History: Rewriting Textbooks in India and the United States." *Georgetown Journal of International Affairs*, Winter /Spring: 101–12.

Vivekananda, Swami. 1992. *The Complete Works of Swami Vivekananda*. Calcutta: Advaita Ashrama.

Voigt, Kevin, Mallika Kapur, and Lonzo Cook. 2013. "Wombs for Rent: India's Surrogate Mother Boomtown." *CNN*, November 2, 2013. www.cnn.com/2013/11/03/world/asia/india-surrogate-mother-industry/.

Vora, Kalindi A. 2013. "Potential, Risk, and Return in Transnational Indian Gestational Surrogacy." *Current Anthropology* 54: S97–106.

———. 2014. "Experimental Sociality and Gestational Surrogacy in the Indian ART Clinic." *Ethnos: Journal of Anthropology* 79: 63–83.

———. 2015a. *Life Support: Biocapital and the New History of Outsourced Labor*. Minneapolis: University of Minnesota Press.

———. 2015b. "Re-imagining Reproduction: Unsettling Metaphors in the History of Imperial Science and Commercial Surrogacy in India." *Somatechnics* 5, no. 1: 88–103.

Wali, Kaeshwar C., ed. 2000. *Satyendra Nath Bose: His Life and Times: Selected Works (with commentary)*. Singapore: World Scientific Publishing Company.

Warren, Karen J. 1990. "The Power and the Promise of Ecological Feminism." *Environmental Ethics* 12, no. 2: 125–46.

Weston, Kath. 2008. "A Political Ecology of 'Unnatural Offences': State Security, Queer Embodiment, and the Environmental Impacts of Prison Migration." *GLQ: A Journal of Lesbian and Gay Studies* 14, no. 2–3: 217–37.

Whitmarsh, Ian, and Elizabeth F. S. Roberts 2016. "Nonsecular Medical Anthropology." *Medical Anthropology: Cross-Cultural Studies in Health and Illness* 35, no. 3: 203–8.

Wilholt, Torsten. 2013. "Epistemic Trust in Science." *British Journal for the Philosophy of Science* 64, no. 2: 233–53.

Willey, Angela. 2016. *Undoing Monogamy: The Politics of Science and the Possibilities of Biology*. Durham, NC: Duke University Press.

Williams, Glynn, and E. Mawdsley. 2006. "Postcolonial Environmental Justice: Government and Governance in India." *Geoforum* 37, no. 5: 660–70.

Williams, Rebecca. 2014. "Storming the Citadels of Poverty: Family Planning under the Emergency in India, 1975–1977." *Journal of Asian Studies* 73, no. 2: 471–92.

Worth, Robert F. 2018. "The Billionaire Yogi Behind Modi's Rise." *New York Times*, July 26, 2018.

Yagnik, Bharat. 2013. "Surrogacy Becomes Family Enterprise in Gujarat." *Times of India*, October 19, 2013. http://timesofindia.indiatimes.com/india/Surrogacy-becomes-family-enterprise-in-Gujarat/articleshow/24365108.cms.

Yap, Audrey Cleo. 2017. "Curriculum Is Biased Against Hinduism." *NBC News*, February 28, 2017.

Zhang, Shulan. 2010. "Conceptualising the Environmentalism in India: Between Social Justice and Deep Ecology." In *Eco-socialism as Politics: Rebuilding the Basis of Our Modern Civilisation*, edited by Huan Qingzhi. Dordrecht: Springer: 181–90.

INDEX

Abraham, Itty, 17, 19, 20, 21, 36
Actor Network Theory, 34, 35, 147, 226
AIDS Bedbhav Virodhi Andolan
  (ABVA), 82
allopathy, 42, 171, 172, 173
Amar Chitra Katha, 3, 219, 237
Anderson, Benedict, 44, 147
Anderson, Warwick, 11, 18, 19, 20, 36
animacy, 73, 107
Anzaldúa, Gloria, v, 29, 30, 209
archaic modernity, 7, 199, 205; definition
  of, 17, 195; description of, 9, 15, 52–53;
  examples of, 21, 23, 68–71, 148, 173,
  200; and masculinity, 54, 57–60, 183
Arondekar, Anjali, 92, 218
ars erotica, 81
Arthashastra, 106
Aryabhata, 242n1
Aryan Migration Theory, 22, 148, 154–64.
  See also Dravidian
Aryans, 149, 162
Arya Samaj movement, 96
Asian values, 21; Asia as method, 20–21.
  See also South Asia
assemblage, 35, 179, 217
avatars: as conceptual tool, 3, 10, 40–42,
  45, 127, 154; as narrative device, 40–
  42, 209, 214, 220; role in mythology,
  40–42, 221
Ayodhya, 56, 114, 116
Ayurveda, 172, 173–74, 199–200, 201;
  Ayurgenomics, 23, 175; comparison
  with other medical systems, 42, 153,

165; products, 14, 15; professional
  degree, 202. *See also* AYUSH
AYUSH, 15, 153, 171, 173, 199, 202

Baba Ramdev, 14–15, 59, 105
Babri Masjid, 56, 57, 116
Bajrang, 57, 117, 239n6
Bamshad, Michael, 152, 154–60
Banerjee, Prathama, 95
Banerjee, Sikata, 10, 54, 95, 99, 116
Basu, Amrita, 54, 55, 56, 57, 59
Benjamin, Ruha, 152, 165, 168, 226
Bharadwaj, Aditya, 17, 196, 197, 200–201,
  215
Bharat, 15, 79, 94, 114, 116; samarth
  Bharat, 200; Swachh Bharat, 136, 170
Bharatiya Janata Party (BJP), 49, 51, 54;
  policies of, 55–62, 105, 119, 192; political platform of, 16, 50, 60–62; in the
  political landscape, 58–60, 68, 69,
  120, 155
bigender, 9
binary: beyond binary, 191, 204, 213; categories of, 54, 87–89, 97; conceptions
  of, 24, 38, 40, 104, 191; logics of, xiii,
  20, 27, 105
biocapital, 168–69
biocitizenship, 145–77
biodiversity, 114, 122
biodivinity, 136, 137
biohumanities, 34, 40
biomoral, 16, 173
bionarrative, 10, 13, 33–40

bionationalism: definition, 8, 10–13, 21–24, 26, 164; ethnic bionationalism, 148, 225, 226; genomic bionationalism, 155, 165, 168–73, 174, 175; Indian bionationalism, 2, 43, 44, 77, 185

biopower, 12, 31, 195

bomb, 58–60, 69, 110; nuclear tests, 58, 59, 60

Brahmin, 94, 132, 228, 220; Brahminical patriarchy, 54, 123, 183; description, 70, 93, 153, 220;

Britain, 13, 17, 48, 79, 84; empire, 13, 81, 84; ideologies, 77, 78, 83

Butalia, Urvashi, 56, 96

California Textbook Controversy, 152–53, 159–62

capital, 16, 167, 169; biocapital, 168, 169; capitalism, 7, 15, 16, 31, 58, 69, 105, 136, 227

caste, 153, 146–54; anticaste, 157, 220; ecological theories, 126–33; and genetics, 146–64; sociology of caste, 131–32; structures, 36, 123, 127, 126–33, 149–50; subcaste, 127, 153, 164, 168. *See also* Dalits

casteism, x, 4, 9, 38, 41, 123, 160; colonialism, 93; politics of, 7, 8, 13, 23, 25; race and caste, 11, 18, 150, 154–59; science of, 10, 11, 18, 22, 99, 134; in social life, 3, 4, 5, 45; women and, 55, 57, 135

celibacy, 93, 94, 95, 99

Chakrabarty, Dipesh, 14, 15, 28, 29, 44, 92

Chakravarti, Uma, 55, 183

Chatterjee, Partha, 54, 93

Christianity, 24–26, 85–85, 90, 93, 97, 148, 220

civilization, 8, 44, 93; logics, 17, 28, 45, 205; Vedic, 6, 15, 79, 184, 216, 242n1

cladogenetic, 132

class, x, 16, 64, 140, 219; biological difference and, 9, 18; politics and, 13, 23,

125, 216; as social category, 36, 126, 189, 190

coconstruction, 9, 51

coloniality, 92

commodification, 21, 189, 190

commodity, 136, 172, 174, 186

communalism, 8, 55, 123

consumerism, 58, 69, 134, 137, 167; biomoral consumerism, 16, 173; consumer class, 125; consumer products, 14, 15, 52, 172, 173, 225

consumption, 225

cyclicality, xvii, 38, 146

Dalits, 153, 196, 244; activism of, 152, 154, 159; Dalit Panthers, 156, 157

Darwin, Charles, 74, 88, 89, 95, 98, 107, 146

Dayananda, Swami, 53, 99

decoloniality, xi, 20, 24, 61, 70, 71, 92

dharma, 26, 38, 135, 136, 146, 183, 217

diaspora, 44, 131, 150–51, 153, 159; diasporic communities, 166, 176, 136; diasporic Indians, 7, 166, 175, 176, 189, 193; diasporic proxies, 166; Hindu diaspora, 43, 162, 226

diversity, xvii, 37, 48, 74, 78, 117, 229; genetic diversity, 166, 168, 169; life, 32, 130; "unity in diversity," x, 164, 174. *See also* biodiversity; Human Genome Diversity Project

DNA, 11, 46, 146, 151, 168, 169; ancestry studies, 152, 154–64; colonial politics, 151–52, 165; indigenous, 10, 166; methods and interpretation, 174–77; mtDNA, 155, 160; Y chromosome, 155, 160

Dravidian, 148, 149, 160, 162

East, xi, 81, 91; superiority of, 7, 54, 118, 127, 183; and West, 20–21, 30, 32, 44, 77, 106, 132, 148, 177, 226

ecology, 11, 61, 116, 118, 125, 136, 139, 225; and caste, 126–33; ecological balance,

122–24; ecological imperialism, 138; econaturalism and primitivism, 134; nationalism, 118, 138–39; native ecologies, 10, 127, 137; sciences, 118, 127, 128, 129, 130, 134, 138; social ecology, 131

effeminization, 55

Egorova, Yulia, 150, 167, 175, 176

emasculation, 55, 100, 102

environment, 65, 140, 167, 170, 218; environmentalism, 22, 113, 213; politics, 22, 68; postcolonial, 113–40, 225

epistemology, 34, 42, 61, 139, 172, 173, 221–22

essentialism, 10, 11, 21, 22, 97, 127, 133, 174

ethno-bionationalism, 225–26

ethnography, 25, 26, 196–97

ethnosciences, 18

eugenics, xii, xiii, 184, 185, 190, 203, 225; classical eugenics, 199; histories of, 78, 98–99, 190; logics of, 13, 187; policies of, 21–22

Eurocentrism, 9, 19, 16, 85

evolution, xvi, xviii, xviii, 5, 6, 46, 74, 81, 146; of caste, 128, 131; debates about, 105, 107; theories of, 28, 29, 89, 132

experimental humanities, 34–40, 41, 221

feminism, 10, 35, 137, 187, 205, 210–11, 225; feminist movements, 183, 187, 191; feminist STS, 10, 34–36, 89, 212, 214, 226; feminist studies, 11, 29, 31, 44, 89, 104, 124, 169, 211; queer feminist STS, 77, 80

feng shui, 63, 64

fertility, 25, 42, 183, 187, 196; hyperfertility, 11, 23, 184, 188; infertility, 225, 241n1

fetus, 41, 188, 204

Foucault, Michel, 12, 31, 81, 186

Gadgil, Madhav, 126–33

Gandhi, Mahatma, 54, 55, 93, 103

Ganesha, 5, 6, 216

garbh sanskar, 23, 200, 201, 203

garbh vigyan sanskar, 202

genetics, 106, 164–74; epigenetics, 41; and genomics 146–64; origin stories, 23; recreational genetics, 151. See also Indian Genome Variation Project (IGVP)

Genographic Projects, 167

genomics: epigenome, 175; genome geographies, 147, 164; genomic nationalism, 148, 155, 164, 165; genomic sovereignty, 23, 152, 165, 176, 226; Indian genomic initiatives, 164–74; national genome, 175

genre, 29, 34, 215–22

globalization, 5, 19, 58, 71, 118, 134, 137, 148, 224, 228; Hinduism and, 176; markets, 16, 147; of reproductive labor, 191

Golwalkar, Madhav, 94, 95, 96

governmentality, 13, 186

great chain of being, 28

green revolution, 53

Guha, Ramachandra, 126–33

Gulf of Mannar, 118, 123

Habib, Irfan, 17

Hanuman, 8, 22, 114, 117–19, 239n6

haplogroups, 155

haplotypes, 155

HapMap(s), 146, 166, 167, 241n1

Haraway, Donna, 27, 33, 34, 36, 37, 80, 130, 140, 187, 212, 215, 216

Harding, Sandra, 18, 19, 20, 27, 212

health, 68, 168, 170, 191, 192, 194. See also public health

Hedgewar, Keshav, 96

helical, 10, 34, 36, 38, 39

heteronormativity, 41, 101, 102, 105

heterosexuality, 41, 87, 89, 93, 204; heterosexual couples, 29, 89; heterosexual families, 192, 195; heterosexual identity, 84, 85, 94, 95; heterosexual marriage, 100; heterosexual sex, 100;

heterosexuality (*continued*)
in Hindu nationalism, 93–96; logics of, 89, 101; in plants and animals, 76
Hindu-centric, 79, 93
Hinduism, 127, 134–5, 176, 192, 220; and diaspora, 159–64; political Hinduism, 7, 8, 10, 11, 42, 96, 205
Hindutva, 7, 14, 16, 51, 62, 151; Hindu nationalism, 50, 56, 204; Hindutva lite, 125
HIV/AIDS, 82, 84, 102–3
homosexuality: and colonialism, 77–78, 80; and Hindu nationalism, 93–97, 105–7; sex and, 14, 22, 81. *See also* Section 377
homosocial, 84, 93, 94
Human Genome Diversity Project, 152, 164, 167

Indian Genome Variation Project (IGVP), 23, 153, 165, 166, 175
Indian Penal Code. *See* Section 377
Indian Space Research Organization (ISRO), 51
indigeneity, 10, 26, 35, 62; knowledge systems, 20, 71, 78; politics, 137–38, 140, 171–74
Indo-Aryan migration, 149; language and, 160
Indus Valley Civilization, 238n4
Islam, 56, 57, 95, 148, 172; histories of, 14; modernism and, 29; pre–Islamic era, 220; science, 35

Jasanoff, Sheila, 175
Judeo–Christian tradition, 26, 37, 78, 84, 137, 213, 218

Kabir (poet–sant), v, x, 38, 220
Kama Sutra, 84, 90, 92
Kapur, Ratna, 80, 82, 84, 96, 97, 104, 189, 190

karma, 26, 38, 43, 146
kinship, 40, 41, 72, 204; genetic kinship, 157
Krishna, Sumi, 123, 129
kshatriya, 93, 94, 153

lingam, xii, 90
Longino, Helen, 212
Lord Macaulay, 80, 241n17
love jihad, 25, 57

Mahabharata, 4, 183, 200, 203, 215; in political life, 6, 79, 102; in popular culture, 5
Majumdar, Partha, 158, 163
Malthus, Thomas, 186
Malthusian logic, 186, 187
Manusmriti, 103, 106
Margulis, Lynn, 107, 164
megadam projects, 61
Menon, Anil, 115, 188, 219, 222
Menon, Nivedita, 56, 118, 119, 121–22, 125
metaphor, 16, 30, 40, 52, 203, 211, 218, 222, 226; Hindu, 11; scientific, 11, 35–36, 41, 132, 145, 204; sexual, 86, 87
militancy, 55, 57
Modi, Narendra: as chief minister of Gujarat, 185, 238n2; development model, 184; development nationalism and, 50, 61, 69, 171; Hindu nationalism and, 7, 8, 40, 41, 60, 133, 199, 204; as prime minister, 5, 6, 21, 49, 106, 116, 137
molecular analyses, 148
molecular biology, 166, 174
molecularization (of life), 26, 167, 173
Moraga, Cherríe, 29
Mughal empire, 91, 148
Mukharji, Projit, 36, 39, 119, 120, 174

Namjoshi, Suniti, 29, 76, 215
Nanda, Meera, 6, 15, 20, 70, 196

Nandy, Ashis, 12, 14, 19, 20, 52, 54, 57, 62, 127
Narmada dam, 60, 134
nativisms, 8, 20, 22, 61, 117–26
naturecultures, 30, 40, 84, 139, 140, 204, 229
Naz Foundation, 80, 82, 103
Nehru, Jawaharlal, x, 15, 16, 49, 52
Neoliberal politics, 22, 23, 61, 135, 136, 137, 145, 177, 191; neoliberalism, 131, 134, 148, 185, 189
nostalgia, 70, 131, 137–40

Omvedt, Gail, 156, 157, 220
1000 Genomes, 146, 167
ontology, 19, 30, 32, 41, 43, 79, 204
Orient, 22, 77, 92, 96; neo–orientalism, 172, 173; oriental, 104, 118; orientalism, 16, 51, 55, 91, 93, 117; orientalist, 53, 91–92, 127, 131, 133, 148–49, 160
origin stories, 33, 142, 146, 147, 148, 151, 163, 220

Pakistan, 56, 58, 60, 147, 155, 220
Pande, Amrita, 185, 189, 190, 198, 199
patriarchy, 9, 56, 225. *See also* Brahmin: Brahminical patriarchy; technopatriarchy
pharmaceutical, 8, 167, 170, 172, 174; pharmaceuticalization, 167; pharmacogenomics, 166
Philip, Kavita, 11, 17, 18, 20, 226
Phule Jyotirao, 220, 221
Pokhran, 58, 59, 60
population, 187, 189
porno–tropics, 91, 92, 93, 104
Prakash, Gyan, 13, 16, 17, 18, 85, 118
Prakriti, 105, 173, 174
Prasad, Amit, 17, 19, 20, 36
primitivity, 28, 48, 95–96, 133
public health, 82, 84, 102–3, 168, 170–71

Puranas, 6, 61, 238n5
Puri, Jyoti, 79, 80, 81, 82, 84, 90, 91, 92, 96, 97, 100

Queer politics, 32, 35, 139, 210, 212; in mythology, 4, 115, 182
Quran, 103, 106

race, 9, 36; biology of, 10, 11, 18, 28–29, 168, 176, 189; and caste, 23, 150, 155–58, 225; and gender, 85, 86; Hindu race, 98, 99; race suicide, 98, 99
Raina, Dhruv, 17, 20
Ramanujan, A. K, 115, 217–20
Ramayana, 4, 113–14, 117; different versions of, 4, 115, 140, 201; politics and, 22, 116, 118–26
Ram Janmabhoomi movement, 56, 116, 117, 118–26
Rashtriya Swayamsevak Sangh (RSS), 11, 105, 202; relation to Hindu nationalism, 50, 54, 93–96
reason, 17, 45, 95, 139, 205, 210; politics and, 24, 33, 40, 130; reason of state, 20, 52, 61; scientific reason, 6, 17
recursivity, xvii, 11, 37, 38
redomestication, 7, 54, 57
relativism, 212
reproduction, 41, 203–4; Indian mythology, 183, 199; labor, 190–92, 194; plants and animals, 78, 86–89; reproductive justice, 98, 191; reproductive technologies, 26, 41, 184, 188–92, 196–97, 204; reproductive tourism, 184
Roberts, Elizabeth, 26, 27, 36, 126, 185, 192, 195, 196, 198, 213
Rudrappa, Sharmila, 188

sacred groves, 123
Sai Baba, 4, 136
Sarkar, Tanika, 54, 55, 58, 96
Savarkar, Vinayak, 96

scheduled castes and tribes, 149, 157, 241n2

Schiebinger, Londa, 18, 29, 86, 87

science, 70, 212; Hindu science, 8, 35, 70; Vedic gestational sciences, 8, 23, 184; Vedic science, 21, 52, 70, 202, 216

Science and Technology Studies (STS), 9, 20, 21, 25, 44, 212; science and religion as binary, 70, 213; secularism and, 26–27; postcolonial STS, 10, 19, 20, 34–36, 212

scientia sexualis, 81

scientize, xi, 9, 10, 31, 67, 189, 199

Section 377, 8, 21; legal history, 78–84, 97–98, 100–107; religion, 24–27, 122, 125; secularism, 9, 176

Shakti, xii, 54, 56, 59, 183

shastras, 62–70, 106, 199, 200

Shiva, x, 11, 54, 84, 90, 182

Shiva, Vandana, 19, 20, 88, 117, 124, 127

Singh, Vandana, 218, 222

South Asia, x, 13, 115, 119, 195, 220; political significance, 15, 79, 96, 148, 216

stratification: reproductive, 189, 192; social, 137, 150, 158, 169

subaltern, 28, 226

Sunder Rajan, Kaushik, 167, 169

Supreme Court, 8, 21, 60. *See also* Section 377

Sur, Abha, 17, 152, 155

surrogacy, 6, 8, 184–86, 188–95, 198–99, 203; altruistic, 195; empowering vs. exploitative, 184, 191–94, 212; medicalized model, 191, 198, 199; surrodev and surro-gods, 198; Surrogacy Regulation Bill 2016, 192, 195

sustainability, 22, 25, 116, 117, 126–33, 135

swadeshi, 15, 153, 164

TallBear, Kim, 152, 226

technology, x, xi, 70, 171, 188; and development, 62–70; and power, 12, 32; and progress, 19–20, 21, 51, 169; science and, xi, 6, 7, 8, 15, 18, 28, 215; technofantasy, 186; technopatriarchy, 184, 186

technoscience(s), 17, 19, 165

temporality, 14, 15, 16, 43

Thackeray, Balasaheb, 58, 59

thigmotropism, 30, 76, 78, 213, 221; as model for narrative, 34–40

Thomas, Renny, 19, 25, 26, 51

time warps, 14, 42

UN World Conference Against Racism, 152, 158

Vaastu/Vastu, 62–70

van der Veer, Peter, 44, 50, 54, 89, 125, 148, 151

Vanita, Ruth, 83, 84

Vishva Hindu Parishad (VHP), 50, 54, 59, 117

vitalism, 31, 32, 45, 207, 213, 214

Vivekananda, Swami, 53, 54

Vora, Kalindi, 185, 186, 189, 190, 199

Western enlightenment, xi, 24, 26, 28, 44, 88, 195; challenges to, xiii, 3, 7, 49, 51; enlightenment positivism, 5, 34; secularism, 195, 198

wombs: in Hinduism xii, 6; for rent, 186, 189. *See also* surrogacy

women: ideal, 53–57, 97–98, 239n9; political organization of, 56–58

worlding, 210, 214, 215

yoga, 7, 11, 14, 90, 136, 199; International Yoga Day, 7. *See also* AYUSH

FEMINIST TECHNOSCIENCES
Rebecca Herzig and Banu Subramaniam, Series Editors

Figuring the Population Bomb: Gender and Demography in the Mid-Twentieth Century, by Carole R. McCann

Risky Bodies & Techno-Intimacy: Reflections on Sexuality, Media, Science, Finance, by Geeta Patel

Reinventing Hoodia: Peoples, Plants, and Patents in South Africa, by Laura A. Foster

Queer Feminist Science Studies: A Reader, edited by Cyd Cipolla, Kristina Gupta, David A. Rubin, and Angela Willey

Gender before Birth: Sex Selection in a Transnational Context, by Rajani Bhatia

Molecular Feminisms: Biology, Becomings, and Life in the Lab, by Deboleena Roy

Holy Science: The Biopolitics of Hindu Nationalism, by Banu Subramaniam